Wilhelm Leutzbach

Introduction to the Theory of Traffic Flow

With 159 Figures

Springer-Verlag Berlin Heidelberg New York
London Paris Tokyo

Professor Dr.-Ing. Wilhelm Leutzbach

Lehrstuhl und Institut für Verkehrswesen
Universität Karlsruhe
Kaiserstrasse 12
D-7500 Karlsruhe/FRG

Extended and totally revised English language version of:
Einführung in die Theorie des Verkehrsflusses.
Springer-Verlag, Berlin Heidelberg New York 1972

ISBN-13: 978-3-642-64805-2   e-ISBN-13: 978-3-642-61353-1
DOI: 10.1007/ 978-3-642-61353-1

Library of Congress Cataloging in Publication Data.
Leutzbach, Wilhelm. Introduction to the theory of traffic flow.
Translation of: Einführung in die Theorie des Verkehrsflusses.
Bibliography: p. Includes index. 1. Traffic flow – Mathematical models. 2. Traffic flow –
Statistical methods. I. Title.
HE336.T7L4813   1988   388.3'143   87-9839

Typesetting: With a system of Springer Produktions-Gesellschaft, Berlin
Dataconversion: Brühlsche Universitätsdruckerei, Giessen
Offsetprinting: H. Heenemann, Berlin. Bookbinding: Lüderitz & Bauer, Berlin
2161/3020-543210

From the Preface to the German Edition

This book describes a coherent approach to the explanation of the movement of individual vehicles or groups of vehicles.

To avoid possible misunderstandings, some preliminary remarks are called for.

1. This is intended to be a textbook. It brings together methods and approaches that are widely distributed throughout the literature and that are therefore difficult to assess. Text citations of sources have been avoided; literature references are listed together at the end of the book.

2. The book is intended primarily for students of engineering. It describes the theoretical background necessary for an understanding of the methods by which links in a road network are designed and dimensioned or by which traffic is controlled; the methods themselves are not dealt with. It may also assist those actually working in such sectors to interpret the results of traffic flow measurements more accurately than has hitherto been the case.

3. The book deals with traffic flow on links between nodes, and not at nodes themselves. Many readers will probably regret this, since nodes are usually the bottlenecks which limit the capacity of the road network. A book dedicated to the node would be the obvious follow-up. A separation of link and node is justified, however, partly because the quantity of material has to be kept within reasonable bounds and partly because the treatment of traffic flow at nodes requires additional mathematical techniques (in particular, those relating to queueing theory).

4. The book presumes a certain level of mathematical knowledge, which should be well within the scope of the engineering graduate. The material could have been dealt with more concisely by taking a purely mathematical approach, but the author wished to make it possible for engineers, for whom mathematics is only a means to an end, to follow derivations through.

5. The treatment of traffic flow is not limited to any particular mode of transport. However, it is clear that the stochastic methods for the description of traffic flow are referred mainly to road traffic.

6. A particular comment should be made on the methods for description of the essentially stochastic, discrete phenomenon of road traffic by means of the deterministic, continuum theory. These are sometimes regarded as relics of a bygone age before stochastic theories of traffic flow were developed. The author does not share this opinion. He is, however, convinced that further development of the theory of traffic flow, particularly any directed at closing existing gaps in the field of partly constrained traffic, will involve stochastic techniques, whether analytical or based on Monte Carlo simulation. Nontheless, on account of the

mathematical difficulties involved, it is still not known whether the potentials of deterministic models have been fully exploited in our search for approximation formulae that can be applied to practical problems.

Preface to the English Edition

The first edition of this book appeared in German in 1972, since when an English edition has been suggested on a number of occasions. Once it had been decided that the book definitely was to appear in English, I took the opportunity to carry out a thorough review of the content and to extend it considerably in parts. For assistance with this I am particularly grateful to Dr. T. Schwerdtfeger.

This edition is based on an earlier translation prepared by Dr. M. B. Godfrey. Dr. M. G. H. Bell has translated the additional material. I am deeply grateful to both of them for their meticulous work. I would also like to thank Dr. P. G. Gipps for linguistic recommendations and Prof. R. E. Allsop for assistance with some of the technical terminology. Responsibility for any errors or omissions remains, of cours, that of the author.

For the arduous task of typing and correcting the manuscript I would like to thank Mrs. B. Lehmann, Miss M. Chimeh and Mr. K. Axhausen, M. S. Thanks are also due to the publisher for devoting so much care and attention to the production of this book.

Karlsruhe, September 1987   W. Leutzbach

Contents

Introduction

It is common to understand the term transportation to mean the change in location of persons, goods, and messages. More specifically let us understand the term traffic flow to mean the change in location of vehicles.

The change in location, or movement, of a vehicle results from the interaction between vehicle and roadway. Here one refers to vehicle dynamics, for which there is a voluminous literature for the various transport modes. This book, however, looks at traffic flow in a different way. The process of a vehicle's motion will be looked at from the point of view of an observer who sees only the motion itself, but not the underlying propulsion process. To the extent that this process of movement, or trajectory, is deterministic, it is dealt with by kinematics; to the extent that is not deterministic, it becomes a problem of mathematical statistics.

Motion will be treated one-dimensionally in a time-distance diagram, even in cases where it occurs in planar or higher-dimensional space. This point of view is preserved even when, for example, longitudinal motion without transverse motion is in practice not thinkable.

Chapter I. The Motion of a Single Vehicle

I.1 Kinematics of a Single Vehicle

I.1.1 Time-dependent Description

I.1.1.1 Motion as a Function of Time

Given any trajectory (Fig. I.1) then, in the time-dependent case:

$x(t)$ is *distance*: as a function of time $[m]$;

$v(t) = \dfrac{dx}{dt}$ is *speed*: as a function of time $[m/s]$;

$b(t) = \dfrac{dv}{dt}$ is *acceleration*: as a function of time

$ = \dfrac{d^2x}{dt^2}$ = the change of speed per unit time $[m/s^2]$;

$k(t)^1 = \dfrac{db}{dt} = \dfrac{d^2v}{dt^2}$ is *jerk*: as a function of time

$ = \dfrac{d^3x}{dt^3}$ = the change of acceleration per unit time $[m/s^3]$.

etc.

If the initial conditions are denoted, respectively, by $t_0[s]$, $x_0[m]$, $v_0[m/s]$, $b_0[m/s^2]$, etc., the following equations of motion result:

$$x(t) = x_0 + \int_{t_0}^{t} v(t)\,dt \tag{I.1}$$

$$v(t) = v_0 + \int_{t_0}^{t} b(t)\,dt \tag{I.2}$$

$$x(t) = x_0 + \int_{t_0}^{t} v_0\,dt + \int_{t_0}^{t}\int_{t_0}^{t} b(t)\,dt\,dt \tag{I.3}$$

---

1 In Chap. II, the traffic density will be denoted by k. Because it is common in kinematics to use k to denote jerk, and because Chap. I does not deal with traffic density, while Chap. II does not use 'jerk', no confusion in the meaning of k should arise.

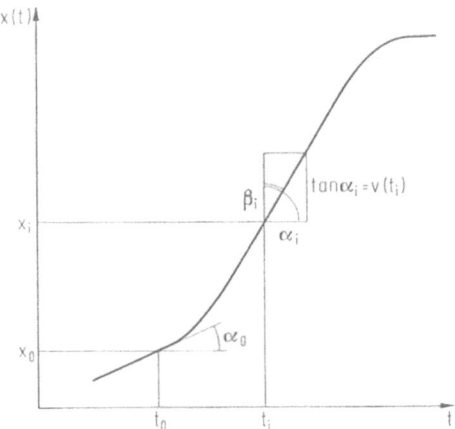

Fig. I.1

$$b(t) = b_0 + \int_{t_0}^{t} k(t)\, dt \tag{I.4}$$

$$v(t) = v_0 \int_{t_0}^{t} b_0 dt + \int_{t_0}^{t} \int_{t_0}^{t} k(t)\, dt\, dt \tag{I.5}$$

$$x(t) = x_0 + \int_{t_0}^{t} v_0 dt + \int_{t_0}^{t}\int_{t_0}^{t} b_0 dt\, dt + \int_{t_0}^{t}\int_{t_0}^{t}\int_{t_0}^{t} k(t)\, dt\, dt\, dt \tag{I.6}$$

$$\vdots$$

etc.

For reasons of simplicity, a value of $k(t) = 0$ will generally be assumed. In the examples which follow for realistic motions, the value of the jerk is, however, very important, because through it are characterized thresholds of comfort.

Example 1. A motion with constant speed is described by

$$b(t) = 0$$

$$v(t) = \text{const}$$

$$x(t) = x_0 + \int_{t_0}^{t} v\, dt = x_0 + v(t - t_0)$$

(see Fig. I.2).[1]

Example 2. For a motion with constant acceleration (decelerations are negative accelerations) (Fig. I.3), we have

$$b(t) = \text{const}$$

$$v(t) = v_0 + \int_{t_0}^{t} b\, dt = v_0 + b(t - t_0)$$

---

1 For the sake of simplicity, here and in the following, this angle will be denoted by v, rather than by arctan v.

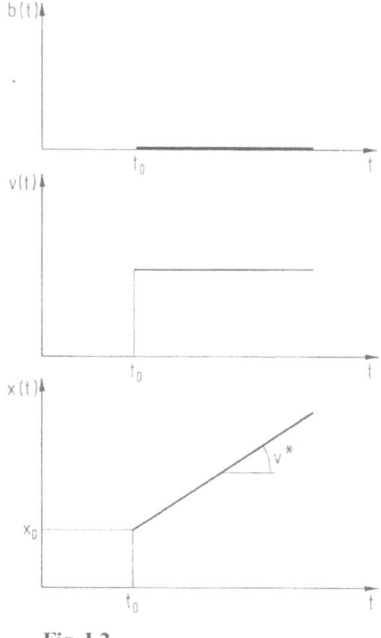

Fig. I.2   Fig. I.3

$$x(t) = x_0 + \int_{t_0}^{t} v(t) \, dt$$

$$= x_0 + \int_{t_0}^{t} [b(t-t_0) + v_0] \, dt$$

$$= x_0 + \frac{1}{2} b(t-t_0)^2 + v_0(t-t_0).$$

Example 3. Consider the braking process of a motor vehicle. In many cases the deceleration is not constant, but, as a first approximation, increases linearly with time. This approximation neglects the response lags and transients of the braking process during the transition from the initial value $b = 0$ to the final value b_0.

For an example calculation of the kinematics of such a process, the following data are given: a vehicle brakes from an initial speed $v_0 = 13.9 \text{ m/s} \, (= 50 \text{ km/h})$ with an initial deceleration $b_0 = -7 \text{ m/s}$ until the vehicle comes to rest. Further, the deceleration is required to reach a final value of $b_e = -9.81 \text{ m/s}$. The braking time and braking distance are to be calculated.

The braking process continues from $t_0 = 0$ to some time t_1. The deceleration is described by

$$b(t) = -(at+c).$$

From the data given, it follows that

$$b(t) = -\left(\frac{2.81}{t_1}t + 7\right).$$

Equation (I.2) states that

$$v(t) = v_0 + \int_{t_0}^{t} b(t)dt.$$

Since $v(t_1) = 0$, we find

$$0 = 13.9 - \int_{t_0}^{t_1} \left(\frac{2.81}{t_1}t + 7\right)dt$$

$$0 = 13.9 - \frac{2.81}{2t_1}t_1^2 - 7t_1.$$

Then, for the braking time, a value of $t_1 = 1.655\,\text{s}$ is obtained.
    Equation (I.3) states that

$$x(t) = x_0 + \int_{t_0}^{t} v_0 dt + \int_{t_0}^{t}\int_{t_0}^{t} b(t)dt\,dt.$$

With $x_0 = 0$, this equation yields the braking distance as

$$x(t_1) = \int_{0}^{t_1} 13.9\,dt - \int_{0}^{t_1}\left(\frac{2.81}{2t_1}t^2 + 7t\right)dt = 12.165\,\text{m}.$$

This distance is, of course, only that distance covered during the actual braking process. Not accounted for is the additional distance covered during the driver's perception and reaction times, during which the vehicle travels at constant speed. Section II.3.3.1.2 goes into more detail on this point.

Example 4. The acceleration values and the corresponding maximum speeds for the individual gears of a vehicle are given in the following table:

Gear	v_{max} [km/h]	b [m/s^2]
1.	30	2.5
2.	60	2.0
3.	150	1.5

The distance travelled by the vehicle and its speed after 15 seconds, when the individual gears are operated up to their corresponding v_{max} values are to be calculated. (The jerk as well as the loss of acceleration due to gear changing is to be ignored.) The calculation has three stages. At the end of a stage:

$$v_E = v_A + b(t_E - t_A)$$

$$t_E = t_A + \frac{v_E - v_A}{b}$$

$$x(t_E) = x(t_A) + v_A \cdot (t_E - t_A) + \frac{1}{2}b(t_E - t_A)^2,$$

where v_A, t_A are the speed and time at the beginning of a stage and v_E, t_E are the speed and time at the end of a stage.

For the first stage in which the driver is using the first gear

$$t_A = 0; \quad x(t_A) = 0; \quad v_A = 0; \quad v_E = 30 \text{ km/h} = 8.33 \text{ m/s}; \quad b_1 = 2.5 \text{ m/s}^2$$

and thus

$$t_E = \frac{8.33}{2.5} = 3.33 \text{ s}$$

$$x(t_E) = 0 + 0 + \frac{1}{2} \cdot 2.5 \cdot 3.33^2 = 13.88 \text{ m}.$$

For the second stage in which the driver uses the second gear

$$t_A = 3.33 \text{ s}; \quad x(t_A) = 13.88 \text{ m}; \quad v_A = 30 \text{ km/h} = 8.33 \text{ m/s};$$

$$v_E = 60 \text{ km/h} = 16.67 \text{ m/s}; \quad b_2 = 2.0 \text{ m/s}^2$$

and therefore

$$t_E = 3.33 + \frac{8.33}{2.0} = 3.33 + 4.17 = 7.50 \text{ s},$$

$$x(t_E) = 13.88 + 8.33(7.50 - 3.33) + \frac{1}{2} \cdot 2.0 \cdot (7.50 - 3.33)^2 = 66.01 \text{ m}.$$

For the third stage in which the driver uses the third gear

$$t_A = 7.50 \text{ s}; \quad x(t_A) = 66.01 \text{ m}; \quad v_A = 60 \text{ km/h} = 16.67 \text{ m/s};$$

$$v_E = 150 \text{ km/h} = 41.67 \text{ m/s}; \quad b_3 = 1.5 \text{ m/s}^2$$

and hence

$$t_E = 7.50 + \frac{25.00}{1.5} = 7.50 + 16.67 = 24.17 \text{ s}.$$

At the end of the third stage, t_E is larger than 15 s, so the vehicle travels only $15 - 7.50 = 7.50$ s in the third gear before the 15 s have elapsed. The distance covered in this period is

$$x(15 \text{ s}) = 66.01 + 16.67 \cdot 7.50 + \frac{1}{2} \cdot 1.5 \cdot 7.50^2 = 233.22 \text{ m}$$

and its speed is then

$$v(15 \text{ s}) = v_A + b_3 \cdot (15 - t_A) = 16.67 + 1.5 \cdot 7.50$$

$$= 27.92 \text{ m/s} = 100.51 \text{ km/h}.$$

Figure I.4 shows the result in graph form.

Example 5. A tram and a car travelling perpendicularly to each other brake simultaneously at time t_0 in order to avoid a collision. Both the tram and car drivers have reaction times of one second. The distance of the tram from the

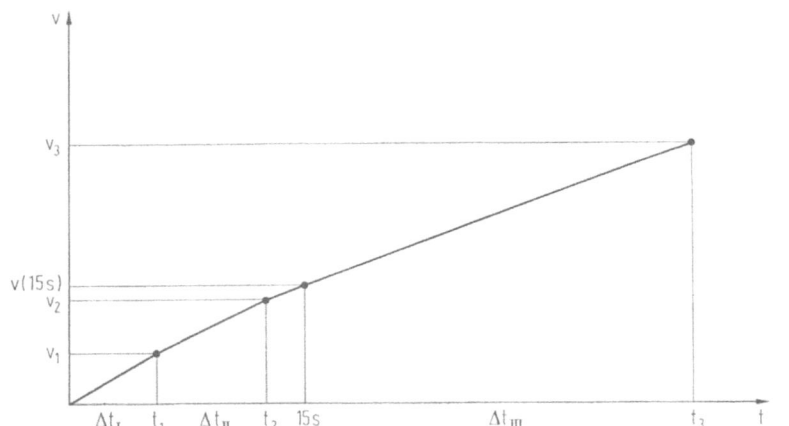

Fig. I.4

potential collision point is 50 m, while that of the car is 30 m. The speed of the tram is $v_A^{tram} = 40$ km/h, and the maximum possible deceleration is $b^{tram} = -2.1$ m/s². The corresponding values for the car are $v_A^{car} = 60$ km/h and $b^{car} = -8.0$ m/s².

It is to be determined whether the two vehicles in fact collide.

The solution is to be found by calculating the stopping distances for the two vehicles, where the stopping distance = reaction distance + braking distance (for comparison see also Sect. II.3.3.1.2).

Initially the reaction distances for the two vehicles are calculated.

The reaction distance x_R is given by

$$x_R = v_A \cdot t_R$$

where v_A is the initial speed and t_R the reaction time. For the car

$$v_A^{car} = 60 \text{ km/h} = 16.67 \text{ m/s}$$

$$x_R^{car} = 16.67 \cdot 1 = 16.67 \text{ m}.$$

For the tram

$$v_A^{tram} = 40 \text{ km/h} = 11.11 \text{ m/s}$$

$$x_R^{tram} = 11.11 \cdot 1 = 11.11 \text{ m}.$$

The braking distances x_B are calculated as follows

$$v_E = v_A + b \cdot t_B; \quad t_B = \frac{v_E - v_A}{b}$$

where v_E is the speed which one wishes to achieve by braking, and t_B is the braking time.

Since braking continues until the vehicles are stationary

$$v_E = 0 \text{ m/s} \quad \text{and} \quad t_B = \frac{-v_A}{b}.$$

The distance covered while braking, $x_B(t_B)$, is in general

$$x_B(t_B) = x_0 + v_A \cdot t_B + \frac{1}{2} bt_B^2.$$

In this case $x_0 = 0\,m$, so

$$x_B(t_B) = v_A \cdot t_B + \frac{1}{2} bt_B^2.$$

Substituting for t_B we obtain

$$x_B(t_B) = v_A \cdot \frac{-v_A}{b} + \frac{1}{2} b \frac{v_A^2}{b^2} = \frac{-v_A^2}{b} + \frac{1}{2} \frac{v_A^2}{b} = -\frac{v_A^2}{2b}.$$

Hence the braking distance for the car is

$$x_B^{car} = -\frac{(v_A^{car})^2}{2 \cdot b^{car}} = -\frac{16.67^2}{2 \cdot (-8)} = 17.37\,m$$

and that for the tram is

$$x_B^{tram} = -\frac{(v_A^{tram})^2}{2 \cdot b^{tram}} = -\frac{11.11^2}{2 \cdot (-2.1)} = 29.39\,m.$$

The stopping distances are

for the car: $16.67 + 17.37 = 34.04\,m$

for the tram: $11.11 + 29.39 = 40.50\,m.$

Since the tram has a stopping distance of 40.50 m but begins to brake at a distance of 50 m from the site of the potential collision, no collision will occur.

Example 6. It is required to calculate the rates of deceleration necessary to stop within 0.5 m when travelling at 36 km/h, 72 km/h, 108 km/h and 144 km/h. This corresponds approximately to a collision with a 0.5 m crumple zone. For comparison, a maximum safe deceleration of $10\,g \simeq 100\,m/s^2$ is assumed for trained astronauts.

Translating the given speeds from [km/h] to [m/s] we obtain

$$v_1 = 36\,km/h = \frac{36}{3.6}\,m/s = 10\,m/s \qquad v_3 = 108\,km/h = \frac{108}{3.6}\,m/s = 30\,m/s$$

$$v_2 = 72\,km/h = \frac{72}{3.6}\,m/s = 20\,m/s \qquad v_4 = 144\,km/h = \frac{144}{3.6}\,m/s = 40\,m/s.$$

The required decelerations are obtained from the braking distance equation

$$x_B = \frac{-v_A^2}{2b} \Rightarrow b = \frac{-v_A^2}{2 \cdot x_B}.$$

For the individual speeds we obtain

$$b_1 = \frac{-10^2}{2 \cdot 0.5 \, m} = -100 \, m/s^2 \simeq -10 \, g \quad b_3 = \frac{-30^2}{2 \cdot 0.5 \, m} = -\ 900 \, m/s^2 \simeq -\ 90 \, g$$

$$b_2 = \frac{-20^2}{2 \cdot 0.5 \, m} = -400 \, m/s^2 \simeq -40 \, g \quad b_4 = \frac{-40^2}{2 \cdot 0.5 \, m} = -1.600 \, m/s^2 \simeq -160 \, g.$$

To calculate the corresponding forces to which the brain is subjected, the following relationship is used:

$$F = m \cdot b$$

where F is the force and m is the mass.

With an assumed brain mass $m = 2 \, kg$, the forces corresponding to the above decelerations are

$$F_1 = 2 \, kg \cdot (100 \, m/s^2) = 200 \, N \qquad F_3 = 2 \, kg \cdot (900 \, m/s^2) \ = 1.800 \, N$$

$$F_2 = 2 \, kg \cdot (400 \, m/s^2) = 800 \, N \qquad F_4 = 2 \, kg \cdot (1.600 \, m/s^2) = 3.200 \, N.$$

I.1.1.2 Motion as a Function of Distance

The equations of motion previously derived are all functions of time. But distance can also be regarded as the independent variable. The ensuing conversion is purely a substitution of variables[1]:

$$v(x) = \frac{1}{dt/dx} \tag{I.7}$$

Write Equation (I.7) in the form

$$\frac{v(x)}{dx} = \frac{1}{dt}$$

and hence

$$dt = \frac{dx}{v(x)}.$$

By integration

$$t(x) = t_0 + \int_{x_0}^{x} \frac{dx}{v(x)}. \tag{I.8}$$

Example 7. For constant speed, $b = 0$ (Fig. I.5), and from Eq. (I.8) one obtains

$$t(x) = t_0 + \frac{x - x_0}{v}.$$

This result would also have followed directly from finding the inverse function of $x(t)$ in Example 1.

---

1 For the sake of clarity, the speed as a function of time, $f(t)$, will henceforth be denoted by $v(t)$, and speed as a function of distance, $g(x)$, will henceforth be denoted by $v(x)$.

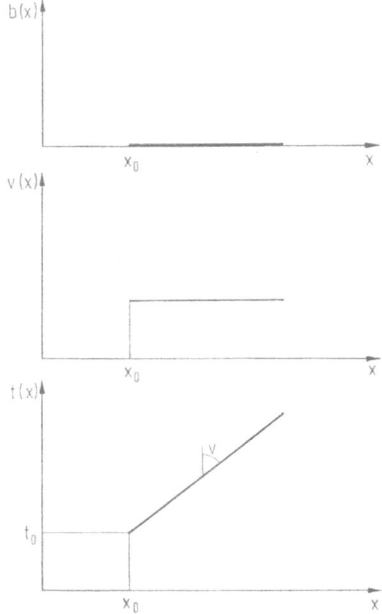

Fig. I.5

Acceleration (which is conditionally defined as a function of time) is obtained as a function of distance $[m/s^2]$ from $v(x)$, with the help of the chain rule

$$b(x) = \frac{d[v(x)]}{dt} = \frac{d[v(x)]}{dt} \cdot \frac{dx}{dx} = \frac{d[v(x)]}{dx} \cdot \frac{dx}{dt}$$

$$= \frac{d[v(x)]}{dx} v(x) = \frac{d\left[\frac{1}{2}(v(x))^2\right]}{dx}. \tag{I.9}$$

This gives

$$d\left[\frac{1}{2}v(x)^2\right] = b(x)\,dx$$

and thence

$$v(x)^2 = v_0^2 + 2\int_{x_0}^{x} b(x)\,dx$$

$$v(x) = \sqrt{v_0^2 + 2\int_{x_0}^{x} b(x)\,dx}. \tag{I.10}$$

Example 8. Consider a motion with $b(t) = b(x) = \text{const}$ (Fig. I.6).
According to Eq. (I.10)

$$v(x) = \sqrt{v_0^2 + 2b(x - x_0)}.$$

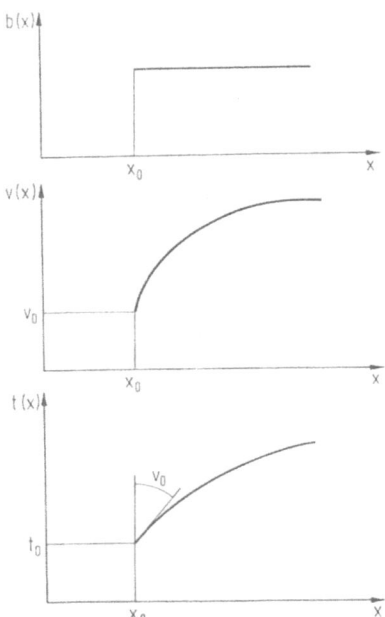

Fig. I.6

$t(x)$ can be calculated in two ways:

1. Using $v(x)$ we have

$$t(x) = \int_{t_0}^{t} dt = t_0 + \int_{x_0}^{x} \frac{dx}{v(x)} = t_0 + \int_{x_0}^{x} \frac{dx}{\sqrt{v_0^2 + 2b(x - x_0)}}.$$

Since it is known in general that

$$\int (a + bx)^n dx = \frac{1}{b(n+1)} (a + bx)^{n+1} + C \quad \text{for } b \neq 0 \text{ and } n \neq -1$$

it follows that

$$t(x) = t_0 + \frac{1}{2b \cdot \frac{1}{2}} \sqrt{v_0^2 + 2b(x - x_0)} \big|_{x_0}^{x} = t_0 - \frac{v_0}{b} + \frac{1}{b} \sqrt{v_0^2 + 2b(x - x_0)}.$$

2. The function $x(t)$ in Example 2

$$x(t) - x_0 = \frac{1}{2} b(t - t_0)^2 + v_0(t - t_0)$$

has the inverse function $t(x)$, where

$$t(x) = t_0 - \frac{v_0}{b} + \frac{1}{b} \sqrt{v_0^2 + 2b(x - x_0)}.$$

With the initial conditions $(t_0, v_0) = 0$, the expression simplifies to

$$t(x) = \frac{1}{b} \sqrt{2b(x - x_0)}.$$

Because t must not become negative, only the positive root is admissible; with the initial condition $v_0 = 0$, b must be positive if any motion at all is to occur. Thus, the following condition must hold

$$\sqrt{v_0^2 + 2b(x - x_0)} \geq 0.$$

I.1.1.3 Motion as a Function of Speed

Consider speed as the independent variable, one can derive

$$b = b(v) = \frac{dv}{dt}; \quad \int_{t_0}^{t} dt = \int_{v_0}^{v} \frac{dv}{b(v)} \quad t(v) = t_0 + \int_{v_0}^{v} \frac{dv}{b(v)}; \tag{I.11}$$

and also

$$b = b(v) = \frac{dv}{dt} = \frac{dv}{dx}\frac{dx}{dt} = \frac{dv}{dx}v = \frac{d\left(\frac{1}{2}v^2\right)}{dx} \quad \int_{x_0}^{x} dx = \int_{v_0}^{v} \frac{v}{b(v)}dv. \tag{I.12}$$

$$x(v) = x_0 + \int_{v_0}^{v} \frac{v}{b(v)}dv. \tag{I.13}$$

Example 9. For the case of constant acceleration

a) $t(v) = t_0 + \dfrac{v - v_0}{b}$

  (which has the inverse function

  $$v(t) = v_0 + b(t - t_0)$$
  as in Example 2), and

b) $x(v) = x_0 + \dfrac{1}{b}\int_{v_0}^{v} v\,dv = x_0 + \dfrac{v^2 - v_0^2}{2b}$

  (which has the inverse function

  $$v(x) = \sqrt{v_0^2 + 2b(x - x_0)}$$
  as in Example 8).

Example 10. If, with a cabin railway, there should be no risk of collision (see Sect. II.3.3.1), the necessary distances between the cabins must be relatively large and consequently the capacity of the links falls. In order to raise this capacity, a reduction of the safety requirements so as to allow collisions with low residual speeds (see Example 38) has been discussed. For a cabin railway with $v_{max} = 10$ m/s and $b_{max} = -4$ m/s^2, we wish to calculate how large the distance between cabins has to be so that, if one cabin suddenly stops, the cabin following collides with a speed of no more than $v_E = 2$ m/s:

$$x(v) = \frac{v_E^2 - v_{max}^2}{2b} = \frac{4 - 100}{-2 \cdot 4} = 12 \text{ m.}$$

Let b be proportional to v,

$$b = av,$$

using Eqs. (I.11) and (I.13) yields

a)

$$t(v) = t_0 + \frac{1}{a} \int_{v_0}^{v} \frac{dv}{v} = t_0 + \frac{1}{a}(\ln v - \ln v_0), \quad v(t) = e^{at+d}$$

(where $d = \ln v_0 - at_0$),

$$x(t) = x_0 + \int_{t_0}^{t} e^{at+d} dt = x_0 + \frac{1}{a}(e^{at+d} - e^{at_0+d}).$$

and b)

$$x(v) = x_0 + \frac{1}{a} \int_{v_0}^{v} dv = x_0 + \frac{v - v_0}{a},$$

$$v(x) = v_0 + a(x - x_0),$$

$$t(x) = t_0 + \int_{x_0}^{x} \frac{dx}{v_0 + a(x - x_0)} = t_0 + \frac{1}{a}\{\ln[v_0 + a(x - x_0)] - \ln v_0\}.$$

In contrast, let b be inversely proportional to v,

$$b = \frac{p}{v}.$$

From Eq. (I.11),

$$t(v) = t_0 + \int_{v_0}^{v} \frac{v\,dv}{p} = t_0 + \frac{v^2 - v_0^2}{2p},$$

$$v(t) = \sqrt{v_0^2 + 2p(t - t_0)},$$

$$x(t) = x_0 + \int_{t_0}^{t} \sqrt{v_0^2 + 2p(t - t_0)}\, dt = x_0 + \frac{2}{3}[v_0^2 + 2p(t - t_0)]^{3/2} \frac{1}{2p}\Big|_{t_0}^{t}$$

$$= x_0 + \frac{1}{3p}\{[v_0^2 + 2p(t - t_0)]^{3/2} - v_0^3\},$$

and from Eq. (I.13)

$$x(v) = x_0 + \int_{v_0}^{v} \frac{v^2}{p} dv = x_0 + \frac{1}{3p}(v^3 - v_0^3),$$

$$v(x) = [3p(x - x_0) + v_0^3]^{1/3},$$

$$t(x) = t_0 + \int_{x_0}^{x} \frac{dx}{v(x)} = t_0 + \int_{x_0}^{x} \frac{dx}{[3p(x - x_0) + v_0^3]^{1/3}}$$

$$= t_0 + \frac{1}{2p}\{[3p(x - x_0) + v_0^3]^{2/3} - v_0^2\}.$$

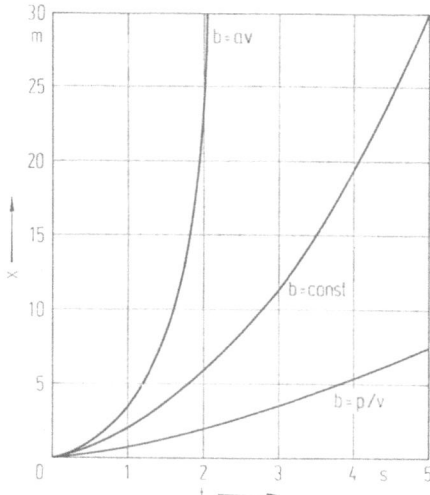

Fig. I.7

Example 11. For the three cases

$$b = \text{const}, \quad b = av, \quad b = p/v$$

with the initial conditions $x_0 = 0$ m, $t_0 = 0$ s, $v_0 = 1$ m/s, $b = 2$ m/s^2, and the parameter values $a = 2s^{-1}$ and $p = 2m^2/s^3$, the three trajectories are depicted in Fig. I.7.

I.1.2 Distance-dependent Description

Even in the preceding discussion when motion was described as a function of distance, speed continued to be, by definition, a function of time, $v = dx/dt$. That led to comparatively unwieldy equations. There is, however, nothing to prevent the description of the same motions in terms of a new parameter which is defined as a function of distance and which is analogous to speed. This means that motion is represented in a t-x-coordinate system, as shown in Fig. I.8.

This new parameter "slowness" = the change in time per unit distance [s/m] as a function of distance is defined as

$$w(x) = \frac{dt(x)}{dx}$$

by analogy with

$$v(t) = \frac{dx(t)}{dt}.$$

Similarly

$$c(x) = \frac{dw(x)}{dx} = \frac{d^2t(x)}{dx^2}$$

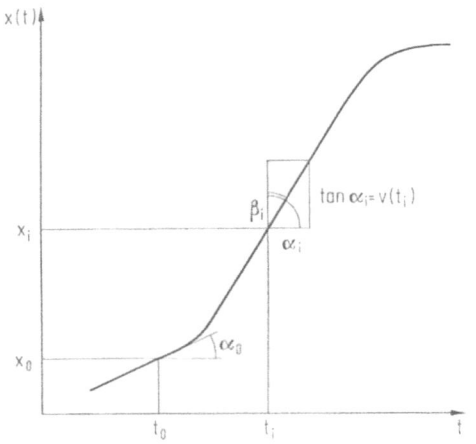 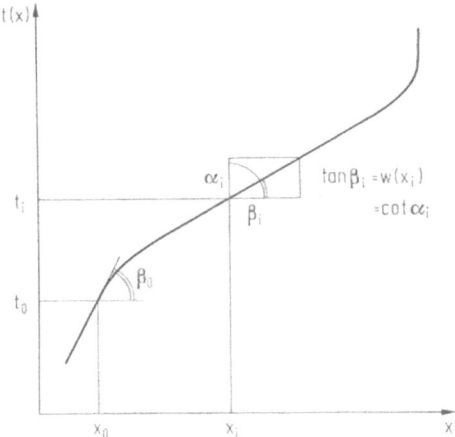

Fig. I.8

by analogy with

$$b(t) = \frac{dv(t)}{dt} = \frac{d^2x(t)}{dt^2}$$

and

$$l(x) = \frac{dc(x)}{dx} = \frac{d^2w(x)}{dx^2} = \frac{d^3t(x)}{dx^3}$$

by analogy with

$$k(t) = \frac{db(t)}{dt} = \frac{d^2v(t)}{dt^2} = \frac{d^3x(t)}{dt^3}$$

etc.

No names are assigned to the parameters $c(x)$ and $l(x)$. The connection between $v(t)$ and $w(x)$ is illustrated in Fig. I.8.

A numerical calculation with w shows one difficulty, that for $v \to 0$, $w \to \infty$ (see Fig. I.9).

As in the preceding sections, equations of motion using w will be developed. Again describing the initial conditions by $t_0[s]$, $x_0[m]$, plus the corresponding inverse variables $w_0[s/m]$, $c_0[s/m^2]$, etc., we have

$$t(x) = t_0 + \int_{x_0}^{x} w(x)\,dx \tag{I.14}$$

$$w(x) = w_0 + \int_{x_0}^{x} c(x)\,dx \tag{I.15}$$

$$t(x) = t_0 + \int_{x_0}^{x} w_0\,dx + \int_{x_0}^{x} \int_{x_0}^{x} c(x)\,dx\,dx \tag{I.16}$$

$$c(x) = c_0 + \int_{x_0}^{x} l(x)\,dx \tag{I.17}$$

$$w(x) = w_0 + \int_{x_0}^{x} c_0\,dx + \int_{x_0}^{x} \int_{x_0}^{x} l(x)\,dx\,dx \tag{I.18}$$

$$t(x) = t_0 + \int\limits_{x_0}^{x} w_0 dx + \int\limits_{x_0}^{x} \int\limits_{x_0}^{x} c_0 dx\, dx + \int\limits_{x_0}^{x} \int\limits_{x_0}^{x} \int\limits_{x_0}^{x} l(x)\, dx\, dx\, dx \qquad (I.19)$$

$$\vdots$$

etc.

Example 12. For motion with $c(x) = 0$, $l(x) = 0$, and $w(x) = $ const, then

$$t(x) = t_0 + \int\limits_{x_0}^{x} w\, dx = t_0 + w(x - x_0)$$

corresponding to $t(x) = t_0 + \dfrac{x - x_0}{v}$ (see Example 7). With $t_0 = 0\,$s and $x - x_0 = 1000\,$m, $t(x) = 1000\,$w, and

$$t/1000 = w,$$

i.e. the time in seconds that a vehicle requires to traverse a distance of 1 km is equal to the slowness w in s/km.

The connection between t, v, and w is made clear by Fig. I.9:

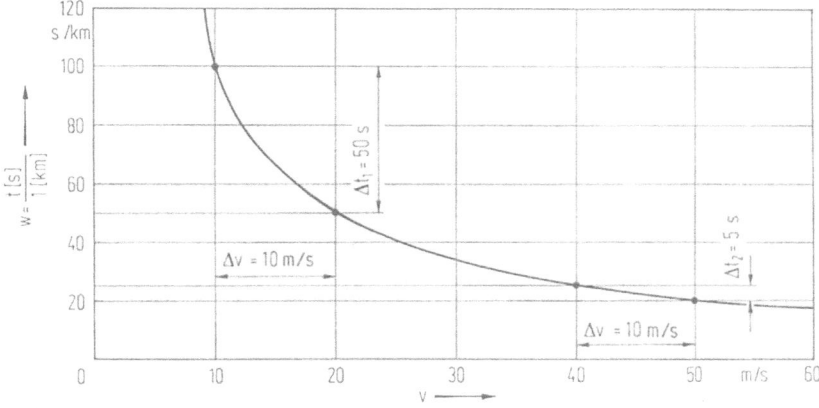

Fig. I.9. (From [6])

The figure shows, in addition, that the savings in travel time (over a distance of 1 km) due to a fixed increase in speed becomes smaller, the greater the value of the final speed.

Example 13. For $l(x) = 0$ and $c(x) = $ const, we have,

$$w(x) = w_0 + \int\limits_{x_0}^{x} c\, dx = w_0 + c(x - x_0)$$

(see Example 2)

$$t(x) = t_0 + \int\limits_{x_0}^{x} w(x)\, dx = t_0 + w_0(x - x_0) + \int\limits_{x_0}^{x} c(x - x_0)\, dx$$

$$= t_0 + w_0(x - x_0) + \frac{1}{2} c(x - x_0)^2.$$

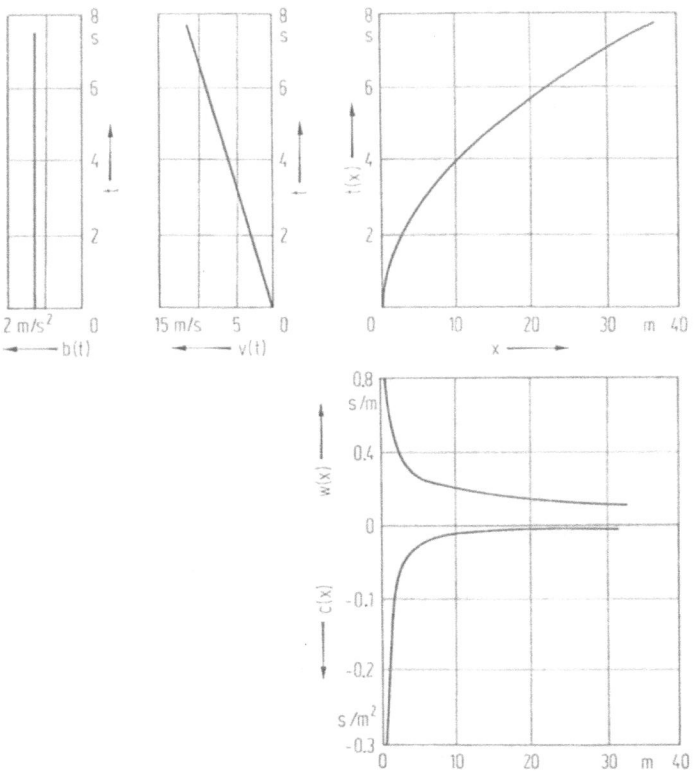

Fig. I.10

Example 14. Figure I.10 illustrates a starting motion for which $(x_0, t_0, v_0) = 0$ and $b = const$, showing $v(t)$ and, in addition, $w(x)$ and $c(x)$:

$$t(x) = \sqrt{\frac{2x}{b}}$$

$$w(x) = \frac{dt}{dx} = \frac{1}{\sqrt{2bx}}$$

$$c(x) = \frac{dw}{dx} = \frac{-1}{2x\sqrt{2bx}} = -\frac{w(x)}{2x}.$$

Example 15. For a trajectory with $c = const$ and the initial conditions $(x_0, t_0, w_0) = 0$, we have by analogy with (Fig. I.11):

$$w(x) = \int_0^x c\, dx = cx; \quad t(x) = \int_0^x cx\, dx = \frac{1}{2}cx^2$$

$$x(t) = \sqrt{\frac{2t}{c}}; \quad v(t) = \frac{1}{\sqrt{2ct}}; \quad b(t) = -\frac{1}{2t\sqrt{2ct}} = -\frac{v(t)}{2t}.$$

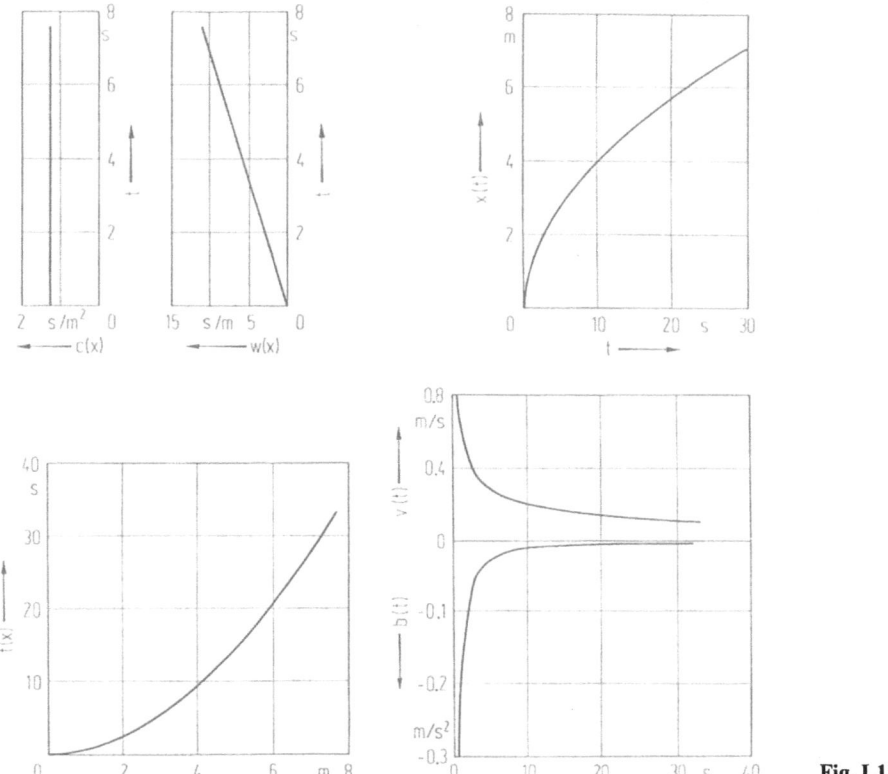

Fig. I.11

Such motion is unrealistic where the initial value of $v(t)$ is infinitely large. It must begin with a positive non-zero value of slowness.

Example 16. Let the initial conditions be $(x_0, t_0) = 0$ and let w_0 and c be constant. From $t(x)$ in Example 13 one obtains the inverse function

$$x(t) = x_0 - \frac{w_0}{c} + \frac{1}{c}\sqrt{2c(t-t_0) + w_0^2}$$

(see Example 8) and, with the initial conditions $(x_0, t_0) = 0$,

$$x(t) = -\frac{w_0}{c} + \frac{1}{c}\sqrt{2ct + w_0^2}.$$

From this, we have

$$v(t) = \frac{2c}{2c\sqrt{2ct + w_0^2}} = \frac{1}{\sqrt{2ct + w_0^2}}.$$

This confirms that, as assumed, the initial slowness w_0 corresponds to the initial speed $v_0 = 1/w_0$, and that, for $t \to \infty$ with $c = \text{const}$, $v(t) \to 0$ (see Fig. I.11). From Eq. (I.10), one obtains for $b(t) = b(x) = \text{const}$ the following

$$w(x) = \frac{1}{v(x)} = \frac{1}{\sqrt{v_0^2 + 2b(x - x_0)}} \tag{I.20}$$

$$c(x) = -\frac{b}{[v_0^2 + 2b(x-x_0)]\sqrt{v_0^2 + 2b(x-x_0)}}.$$ (I.21)

Subsequently, we find

a) it is not possible for both $c(x)$ and b to be constant

and

b) $c(x)$ and b have opposite signs because the denominator of Eq.(I.21) cannot become negative (see Example 8, part 2).

I.1.3 Graphical Transformations

In cases where the variables describing the motion, as a function of either time or distance, are given in a form which renders an analytical description difficult or

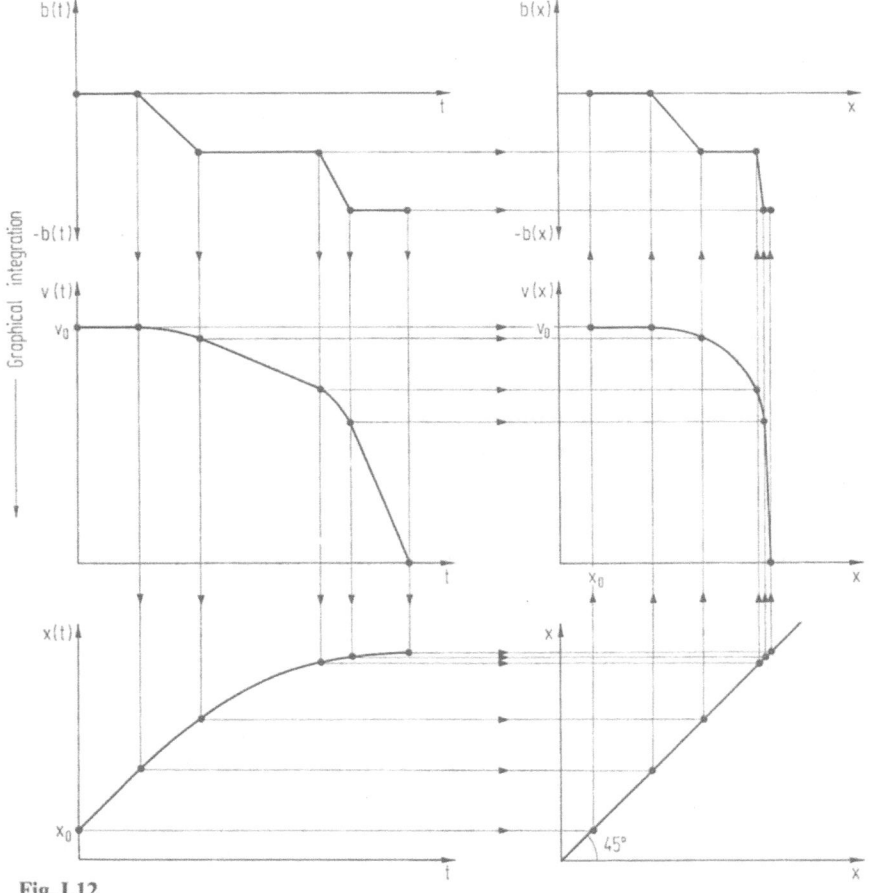

Fig. I.12

impossible, one can use graphical differentiation to compute $v(t)$ and $b(t)$ from a given $x(t)$, and use graphical integration to obtain $v(t)$ and $x(t)$ from a given function $t(x)$, $v(x)$, or $b(x)$.

When motion is described as a function of time, one can transform it into a function of distance, (i.e. find the inverse function), this can be done graphically by reflection about a 45°-line (Fig. I.12).

I.2 Statistics of the Motion of an Individual Vehicle

In many practical cases the equations of motion described in the preceding sections are a useful approximation. In reality we frequently find fluctuations in speed and acceleration so that the motion gives the impression of being more random than deterministic. This is especially true for motor vehicle traffic in which the motion of a particular vehicle at a particular moment is not only dictated by the will of the driver (which itself fluctuates from moment to moment) but is also strongly influenced by road conditions, weather conditions, the presence of other vehicles on the road etc. Similarly, ships and aircraft are subject to speed fluctuations through, for example, the variations in water currents and in atmospheric conditions. Figure I.13 illustrates such a motion in a time-distance diagram.

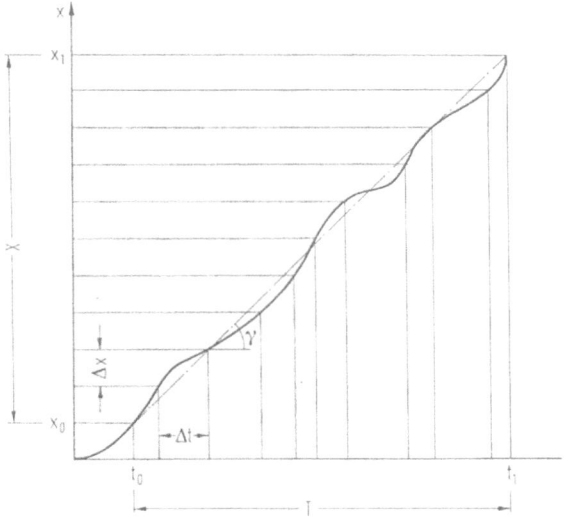

Fig. I.13

I.2.1 Means and Variances of Speeds

The description of such irregular motions is possible statistically. Had one, for example, measured the speed during m time intervals, or over n distance intervals, and recorded these measurements on a speed histogram with k classes, there being m_i or n_i entries in class i, covering the interval

$$\left(\text{by} \quad \sum_{i=1}^{k} m_i = m \quad \text{or} \quad \sum_{i=1}^{k} n_i = n \right)$$

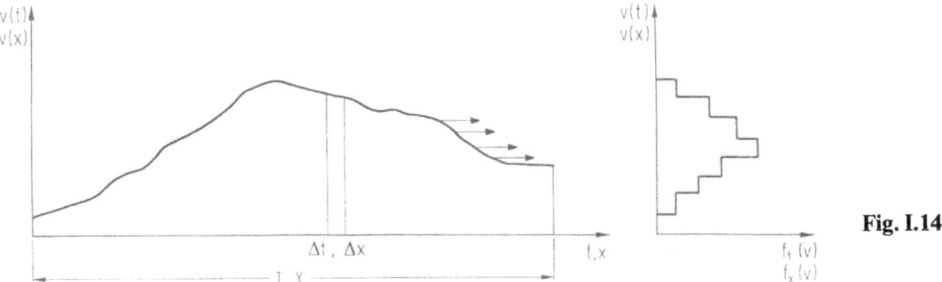

Fig. I.14

one would obtain the absolute frequency distribution of the speed of an observed vehicle either during the time T or over distance X as shown in Fig. I.14. The speed record over time will be called the speed-time profile; the record of the speed over distance will be called the speed-distance profile. Instead of the actual frequencies m_i or n_i, we use the quotient of the absolute frequency, dividing by m or n:

$$m_i/m = f_t(v_i), \quad \text{if v is a function of time, and}$$

$$n_i/n = f_x(v_i), \quad \text{if v is a function of distance}$$

thereby obtaining the relative frequency, and through summation the relative cumulative frequency. In order to describe an empirical frequency distribution numerically, various statistical quantities will suffice in general: the arithmetic mean, and the variance, or mean square deviation from the arithmetic mean.

Example 17. Suppose that speeds have been measured in m = 229 intervals of a speed-time profile (for example with a tachograph). These speeds could then be summarized as follows:

Classes of speed	Absolute frequency of time intervals	Relative frequency $f_t(v_i)$	Relative cumulative frequency $F_t(v \leq v_i)$
22.5–27.5	3	0.0132	0.0132
27.5–32.5	5	0.0218	0.0350
32.5–37.5	10	0.0436	0.0786
37.5–42.5	22	0.0961	0.1747
42.5–47.5	31	0.1354	0.3101
47.5–52.5	33	0.1440	0.4541
52.5–57.5	35	0.1528	0.6069
57.5–62.5	31	0.1354	0.7423
62.5–67.5	23	0.1006	0.8429
67.5–72.5	19	0.0830	0.9259
72.5–77.5	10	0.0436	0.9695
77.5–82.5	5	0.0218	0.9913
82.5–87.5	2	0.0087	1.0000
	m = 229	1.000	

Mean:

$$\bar{v}_t = \frac{1}{m} \sum_{i=1}^{m} v_i = \frac{1}{229}(25 + 25 + 25 + 30 + 30 + 30 + 30 + \ldots + 85 + 85)$$

$$= 54.0 \text{ km/h.}$$

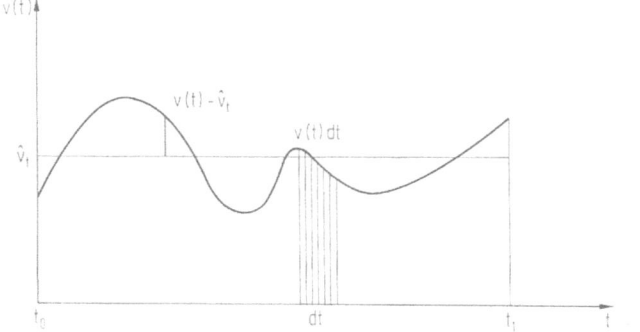

Fig. I.15

No account is taken of the distribution of the actual values of the speeds within each class interval, the midpoint being used instead.

Variance:

$$s_t^2 = \frac{1}{m-1} \sum_{i=1}^{m} (v_i - \bar{v}_t)^2 = 151 \text{ km}^2/\text{h}^2.$$

(In answer to the question, why $m-1$ instead of m, the reader is referred to the relevant statistical literature.)

Standard deviation (the square root of the variance):

$$s_t = \sqrt{s^2} = 12.3 \text{ km/h.}$$

The mean value represents the height of a rectangle whose area is the same as the area under the speed-time or -distance profile, into a rectangle of equal area (see Fig. I.15). Thus, with

$$\int_{t_0}^{t_1} dt = t_1 - t_0 = T; \quad \hat{v}_t = \frac{\int_{t_0}^{t_1} v(t)\,dt}{\int_{t_0}^{t_1} dt} = \frac{1}{T} \int_{t_0}^{t_1} v(t)\,dt \tag{I.22}$$

resp. and with

$$\int_{x_0}^{x_1} dx = x_1 - x_0 = X; \quad \hat{v}_x = \frac{\int_{x_0}^{x_1} v(x)\,dx}{\int_{x_0}^{x_1} dx} = \frac{1}{X} \int_{x_0}^{x_1} v(x)\,dx, \tag{I.23}$$

$\hat{v}_t$, the mean value of the speed-time profile, will be referred to as the journey speed, $\hat{v}_x$, the mean value of the speed profile, will be referred to as the route speed. Because in Eq. (I.22)

$$\int_{t_0}^{t_1} v(t)\,dt = \int_{t_0}^{t_1} \frac{dx}{dt}\,dt = \int_{x_0}^{x_1} dx = X$$

the journey speed corresponds to the slope of a straight line between the points (t_0, x_0) and (t_1, x_1): in the time-distance diagram,

$$\hat{v}_t = \text{tg } x = X/T$$

(see Fig. I.13).

As in Example 17, the variance of a speed-time profile or a speed-distance profile is calculated as the mean square difference from the mean (see Example I.15):

$$\sigma_t^2 = \frac{1}{T} \int\limits^T [v(t) - \hat{v}_t]^2 dt = \frac{1}{T} \left[\int\limits^T v(t)^2 dt - 2\hat{v}_t \int\limits^T v(t) dt + \hat{v}_t^2 \int\limits^T dt \right]$$

or, with

$$\frac{1}{T} \int\limits^T v(t) dt = \hat{v}_t \quad \text{and} \quad \int\limits^T dt = T$$

$$\sigma_t^2 = \frac{1}{T} \int\limits^T v(t)^2 dt - \hat{v}_t^2. \tag{I.24}$$

Correspondingly,

$$\sigma_x^2 = \frac{1}{X} \int\limits^X v(x)^2 dx - \hat{v}_x^2 \tag{I.25}$$

Example 18. A vehicle moves between two points A and B with constant acceleration b from $v = 0$ until $v = v_{max}$. As a function of time

$$v = v(t) = bt$$

$$\hat{v}_t = \frac{1}{T} \int\limits^T bt \, dt = \frac{1}{2} v_{max},$$

which one obtains directly by transforming the area of the triangle formed by the speed-time profile into a rectangle of equal area on the same base.
The variance is

$$\sigma_t^2 = \frac{1}{T} \int\limits^T b^2 t^2 dt - \left(\frac{1}{2} v_{max}\right)^2 = \frac{1}{12} v_{max}^2.$$

By similar calculations, one finds, that

$$v(x) = \sqrt{2bx}$$

$$\hat{v}_x = \frac{1}{X} \int\limits^X \sqrt{2bx} \, dx = \frac{2}{3} v_{max}$$

$$\sigma_x^2 = \frac{1}{X} \int\limits^X 2bx \, dx - \left(\frac{2}{3} v_{max}\right)^2 = \frac{1}{18} v_{max}^2.$$

(Example 18 will be continued on pp. 29 and 30.)

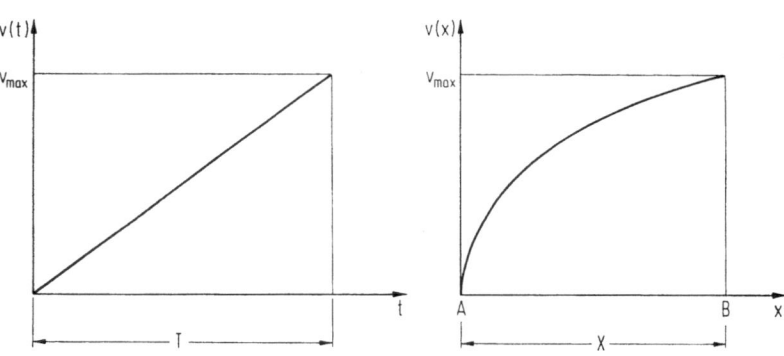

Fig. I.16

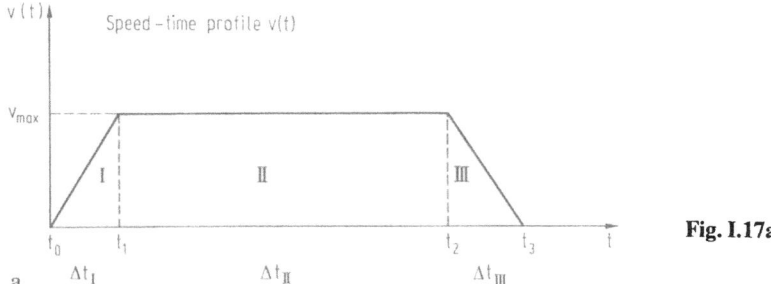

$v(t)$

Speed–time profile $v(t)$

v_{max}

I

II

III

t_0 t_1 t_2 t_3 t

a Δt_I Δt_{II} Δt_{III}

Fig. I.17a

Example 19. A tram has constant acceleration $b = 1.2 \text{ m/s}^2$ up to its maximum speed $v_{max} = 36 \text{ km/h} = 10 \text{ m/s}$. From a tram stop, the tram accelerates at $+b$ to achieve its maximum speed, which it maintains for 50 seconds before decelerating at $-b$ until it comes to rest again. We wish to calculate:

a) the journey speed $\hat{v}_t$,
b) the distance covered before maximum speed is attained, and
c) the route speed $\hat{v}_x$.

Regarding a), the situation can be clarified with the assistance of a speed-time profile (Fig. I.17a).

In Fig. I.17a the movement of the tram is divided into three phases:

Phase I   : Acceleration from rest to v_{max}
Phase II : Travelling at v_{max}
Phase III : Braking from v_{max} to rest.

The journey speed $\hat{v}_t$ can be calculated from Eq. (I.22)

$$\hat{v}_t = \frac{1}{T} \int_T v(t)\,dt.$$

The calculation has three parts corresponding to the phases:

$$\int_T v(t)\,dt = \int_{t_0}^{t_1} v(t)\,dt + \int_{t_1}^{t_2} v(t)\,dt + \int_{t_2}^{t_3} v(t)\,dt.$$

For Phase I

$$v_I(t) = v_0 + b\cdot(t - t_0); \quad x_I(t) = x_0 + \frac{1}{2}\cdot bt^2.$$

Setting $v_0 = 0 \text{ m/s}$ and $t_0 = 0$ yields

$$v_{max} = b\cdot\Delta t_I \rightarrow \Delta t_I = \frac{v_{max}}{b}$$

$$\Delta t_I = \frac{10 \text{ m/s}}{1.2 \text{ m/s}^2} = 8.33 \text{ s}$$

and generally

$$v_I(t) = b\cdot t.$$

For Phase II, $v_{max} = v_{II} = 10$ m/s and $\Delta t_{II} = 50$ s.

The calculation for Phase III follows that for Phase I. Since $v_3 = v_{max} - b\Delta t_{III}$, where $v_3 = 0$ m/s is the terminal speed, we obtain:

$$\Delta t_{III} = \frac{-v_{max}}{-b} = \frac{-10\,\text{m/s}}{-1.2\,\text{m/s}^2} = 8.33\text{ s}$$

and generally

$$v_{III}(t) = v_{max} - b \cdot t.$$

The individual times are now:

$$
\begin{aligned}
t_0 &= 0\text{ s}\\
t_1 &= t_0 + \Delta t_I &&= 0 &&+ 8.33 &&= 8.33\text{ s}\\
t_2 &= t_1 + \Delta t_{II} &&= 8.33 &&+ 50 &&= 58.33\text{ s}\\
t_3 &= t_2 + \Delta t_{III} &&= 58.33 &&+ 8.33 &&= 66.66\text{ s} = T.
\end{aligned}
$$

In this way, Eq. (I.22) can be evaluated. By inserting $v_I(t)$, $v_{II}(t)$ and $v_{III}(t)$

$$\hat{v}_t = \frac{1}{T}\left[\int_{t_0}^{t_1} v_I(t)\,dt + \int_{t_1}^{t_2} v_{II}(t)\,dt + \int_{t_2}^{t_3} v_{III}(t)\,dt \right]$$

and hence

$$\hat{v}_t = \frac{1}{T}\left[\int_{t_0}^{t_1} b \cdot t \cdot dt + \int_{t_1}^{t_2} v_{max} \cdot dt + \int_{t_2}^{t_3} (v_{max} - b \cdot t) \cdot dt \right].$$

Following integration:

$$
\begin{aligned}
\hat{v}_t &= \frac{1}{T}\left[\frac{1}{2} \cdot b \cdot t^2 \big|_{t_0}^{t_1} + v_{max} \cdot t \big|_{t_1}^{t_2} + v_{max} \cdot t - \frac{1}{2} \cdot b \cdot t^2 \big|_{t_2}^{t_3} \right]\\[2mm]
&= \frac{1}{T}\left\{ \left[\frac{1}{2} \cdot b(t_1 - t_0)^2 \right] + v_{max}(t_2 - t_1) + v_{max}(t_3 - t_2) \right.\\[2mm]
&\quad \left. - \left[\frac{1}{2} \cdot b(t_3 - t_2)^2 \right] \right\}\\[2mm]
&= \frac{1}{66.66}\left\{ \left[\frac{1}{2} \cdot 1.2 \cdot (8.33 - 0)^2 \right] + 10(58.33 - 8.33) + 10(66.66 - 58.33) \right.\\[2mm]
&\quad \left. - \left[\frac{1}{2} \cdot 1.2 \cdot (66.66 - 58.33)^2 \right] \right\}\\[2mm]
&= \frac{1}{66.66}[41.63 + 500 + 83.30 - 41.63]\\[2mm]
&= 8.75\text{ m/s} \cong 31.5\text{ km/h.}
\end{aligned}
$$

Regarding b), the distance covered in Phase I is calculated from

$$x_1(t) = x_0 + v_0(t - t_0) + \frac{1}{2}b(t - t_0)^2$$

where $x_0 = 0$, $v_0 = v$, $t_0 = t$ and $t = t_1$ yield

$$x_I(t_1) = \frac{1}{2} \cdot 1.2 \cdot 8.33^2 = 41.63 \text{ m}.$$

Regarding c), the route speed is given by

$$\hat{v}_x = \frac{1}{X} \int\limits_X v(x)\,dx$$

where X is the total distance covered.

Calculation is based on the speed-distance profile $v(x)$ and follows the three stages identified earlier

$$\int\limits_X v(x)\,dx = \underbrace{\int\limits_{x_0}^{x_1} v_I(x)\,dx}_{\text{Phase I}} + \underbrace{\int\limits_{x_1}^{x_2} v_{II}(x)\,dx}_{\text{Phase II}} + \underbrace{\int\limits_{x_2}^{x_3} v_{III}(x)\,dx}_{\text{Phase III}}$$

First $x_i (i = 1,2,3)$ is determined.

From the general equation

$$v(x) = \sqrt{v_0^2 + 2 \cdot b(x - x_0)} \quad (\text{see Example 9})$$

and setting $v_0 = 0$ m/s and $x_0 = 0$ m we obtain

$$v_I(x) = \sqrt{2bx}; \quad v_I(x_1) = \sqrt{2 \cdot 1.2 \cdot x_1} = 10 \text{ m/s}$$

and thus

$$x_1 = 41.63 \text{ m}.$$

For Phase II

$$v_{II}(x) = v_{max} = \text{const} = 10 \text{ m/s}$$

$$\Delta x_{II} = v_{max}\Delta t_{II} = 10 \cdot 50 = 500 \text{ m}$$

$$x_2 = x_1 + \Delta x_{II} = 41.63 + 500 = 541.63 \text{ m}.$$

The distance covered in Phase III is the same as that in Phase I:

$$\Delta x_{II} = 41.63 \text{ m}$$

$$x_3 = x_2 + \Delta x_{III} = 541.63 + 41.63 = 583.26 \text{ m} = X$$

and hence

$$v_{III}(x) = \sqrt{v_{max}^2 + 2b(x - x_2)}.$$

In this way

$$\hat{v}_x = \frac{1}{X}\left[\int\limits_{x_0}^{x_1} v_I(x)\,dx + \int\limits_{x_1}^{x_2} v_{II}(x)\,dx + \int\limits_{x_2}^{x_3} v_{III}(x)\,dx\right].$$

Inserting $v_I(x)$, $v_{II}(x)$, $v_{III}(x)$:

$$\hat{v}_x = \frac{1}{X}\left[\int\limits_{x_0}^{x_1} \sqrt{2bx} \cdot dx + \int\limits_{x_1}^{x_2} 10\,dx + \int\limits_{x_2}^{x_3} \sqrt{v_{max}^2 + 2 \cdot b(x - x_2)} \cdot dx\right].$$

After integration:

$$\hat{v}_x = \frac{1}{X}\left[\sqrt{2b}\cdot\frac{2}{3}x^{3/2}\Big|_{x_0}^{x_1} + 10x\Big|_{x_1}^{x_2} + \frac{(v_{max}^2-2bx_2+2bx)^{3/2}}{3b}\Big|_{x_2}^{x_3}\right]$$

$$= \frac{1}{X}\left[\sqrt{2b}\left(\frac{2}{3}x_1^{3/2}-\frac{2}{3}x_0^{3/2}\right)+10(x_2-x_1)\right.$$

$$\left.+\frac{(v_{max}^2-2bx_2+2bx_3)^{3/2}-(v_{max}^2-2bx_2+2bx_2)^{3/2}}{3b}\right]$$

$$= \frac{1}{583.26}\left[\sqrt{2\cdot1.2}\left(\frac{2}{3}\cdot41.63^{3/2}-0\right)+10(541.63-41.63)\right.$$

$$\left.+\frac{[100-2(-1.2)541.63+2(-1.2)583.26]^{3/2}-[100-2(-1.2)541.63+2(-1.2)541.63]^{3/2}}{3(-1,2)}\right.$$

$$= \frac{1}{583.26}(277.41+5000+277.77)$$

$$= 9.52\,\text{m/s} \cong 34.27\,\text{km/h}.$$

Speed-distance profile, (Fig. I.17b).

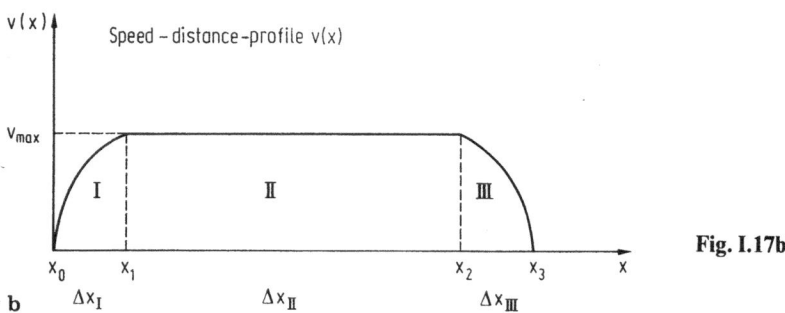

Fig. I.17b

If the speed of a vehicle is given as a function of time or distance, an empirical density function and an empirical distribution function can be computed in analogy to a probability density function and a cumulative distribution function[1]: The frequency of appearance of a particular speed v_i is that amount of time, Δt_i (or of distance, Δx_i), during which the speed has the measured value v_i as in relation to the entire observation time T (or the observation distance X):

$$f_t(v_i) = \frac{\Delta t_i}{T} \quad\text{or}\quad f_x(v_i) = \frac{\Delta x_i}{X}. \tag{I.26}$$

The cumulative distribution function is obtained through summation:

$$F_t(v\leqq v_i) = \sum_{v\leqq v_i}\frac{\Delta t_k}{T} \quad\text{or}\quad F_x(v\leqq v_i) = \sum_{v\leqq v_i}\frac{\Delta x_k}{X}. \tag{I.27}$$

When the intervals Δt and Δx are small enough, the summation in Eq.(I.27) can be changed to integrations

---

1 For the reader not yet familiar with mathematical statistics, it is recommended that Sect. II.1 be consulted.

$$F_t(v) = \frac{1}{T} \int\limits_0^v \frac{dt}{dv} dv \quad \text{or} \quad F_x(v) = \frac{1}{X} \int\limits_0^v \frac{dx}{dv} dv. \tag{I.28}$$

The quotient dt/dv, or dx/dv, is just the derivative of the inverse function of $v(t)$ or $v(x)$. Because one can write the empirical density function as

$$dF_t(v) = f_t(v) dv; \quad dF_x(v) = f_x(v) dv, \tag{I.29}$$

$$f_t(v) = \frac{1}{T} \frac{dt}{dv}; \quad f_x(v) = \frac{1}{X} \frac{dx}{dv}.$$

Most importantly

$$\int\limits_0^{v_{max}} f_t(v) dv = 1 \quad \text{or} \quad \int\limits_0^{v_{max}} f_x(v) dv = 1. \tag{I.29a}$$

In contrast to the procedure followed in Example 14, the mean and variance can be calculated using the density function:

$$\bar{v}_t = \int\limits_0^{v_{max}} v f_t(v) dv; \qquad \bar{v}_x = \int\limits_0^{v_{max}} v f_x(v) dv^2 \tag{I.30}$$

$$\bar{\sigma}_t^2 = \int\limits_0^{v_{max}} (v-\bar{v}_t)^2 f_t(v) dv; \qquad \bar{\sigma}_x^2 = \int\limits_0^{v_{max}} (v-\bar{v}_x)^2 f_x(v) dv. \tag{I.31}$$

The quantities calculated with Eqs. (I.30) and (I.31) must be identical with those derived from the speed-time and the speed-distance profiles.

Example 18 (continued). For the speed-time-profile with $b = \text{const}$, we have, with reference to Eq. (I.29), the inverse function $t(v) = (1/b)v$ and $dt/dv = 1/b$

$$f_t(v) = \frac{1}{bT}$$

or, with $v_{max} = bT$

$$f_t(v) = \frac{1}{v_{max}}$$

(see Fig. I.18 a).
From Eq. (I.30) we have

$$\bar{v}_t = \int\limits_0^{v_{max}} v \frac{1}{v_{max}} dv = \frac{1}{2} v_{max}$$

---

1 The upper limit of integration v_{max} means that v must be a monotone increasing function. Otherwise, a piecewise calculation is required.
2 In order to differentiate between means and variances calculated from density functions, speed-time or -distance profiles, the following notation is used: calculated from

speed-time profiles $v(t) = \hat{v}_t, \sigma_t^2$ density function $f_t(v) = \bar{v}_t, \bar{\sigma}_t^2$

speed-distance profiles $v(x) = \hat{v}_x, \sigma_t^2$ density function $f_x(v) = \bar{v}_x, \bar{\sigma}_x^2$

as above, and from Eq. (I.31)

$$\bar{\sigma}_t^2 = \int_0^{v_{max}} (v - \bar{v}_t)^2 \frac{1}{v_{max}} dv = \int_0^{v_{max}} v^2 \frac{1}{v_{max}} dv - \left(\frac{1}{2} v_{max}\right)^2 = \frac{1}{12} v_{max}^2$$

also as above.

I.2.2 The Relationship Between the Parameters of Time-dependent and Distance-dependent Motion

It will be shown below that there is a mathematical relationship between the density function of the speed measured over time $f_t(v)$, and the density function over distance, $f_x(v)$ [cf. also Sect. II.2.3.2, Eq. (II.26)]:

If a vehicle traverses a distance X in a time T, the duration of the time interval during which it is travelling at speed v is

$$t_v = Tf_t(v).$$

The actual distance over which it is travelling at speed v is

$$x_v = Xf_x(v).$$

But x_v and t_v are related, since

$$x_v = vt_v.$$

Thus, we obtain,

$$Xf_x(v) = Tf_t(v)v; \quad f_x(v) = \frac{v}{X/T} f_t(v)$$

and, because $X/T = \bar{v}_t$

$$f_x(v) = \frac{v}{\bar{v}_t} f_t(v). \tag{I.32}$$

Example 18 (continued). Now we can write

$$f_x(v) = v \frac{2}{v_{max}} \cdot \frac{1}{v_{max}} = \frac{2v}{v_{max}^2}$$

(see Fig. I.18b) which can also be obtained from Eq. (I.29), and Eq. (I.30). We have:

$$\bar{v}_x = \int_0^{v_{max}} \frac{2v^2}{v_{max}^2} dv = \frac{2}{3} v_{max}$$

and from Eq. (I.31)

$$\bar{\sigma}_x^2 = \int_0^{v_{max}} \frac{2v^3}{v_{max}^2} dv - \left(\frac{2}{3} v_{max}\right)^2 = \frac{1}{18} v_{max}^2.$$

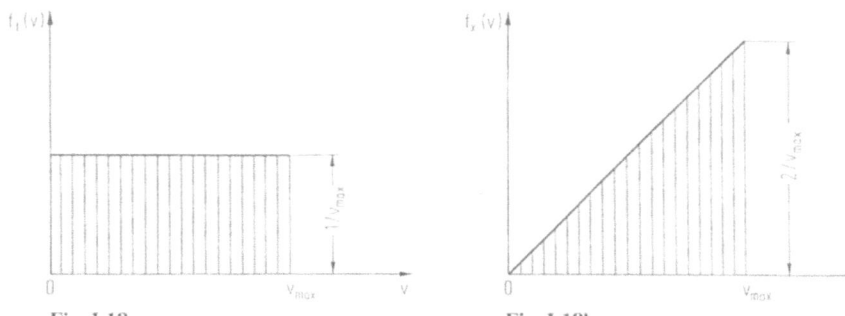

Fig. I.18a   Fig. I.18b

Therefore, there is no difference if one computes the desired statistical measures from the speed-time or -distance profiles, or from the respective empirical density functions.

The relationship between v_x and v_t can be calculated using Eq.(I.32)

$$\bar{v}_x = \int_0^{v_{max}} v f_x(v)\,dv = \int_0^{v_{max}} \frac{v^2}{\bar{v}_t} f_t(v)\,dv = \frac{1}{\bar{v}_t} \int_0^{v_{max}} v^2 f_t(v)\,dv. \tag{I.33}$$

From Eq.(I.31) we obtain

$$\bar{\sigma}_t^2 = \int_0^{v_{max}} (v - \bar{v}_t)^2 f_t(v)\,dv = \int_0^{v_{max}} (v^2 - 2v\bar{v}_t + \bar{v}_t^2) f_t(v)\,dv$$

$$= \int_0^{v_{max}} v^2 f_t(v)\,dv - 2\bar{v}_t \int_0^{v_{max}} v f_t(v)\,dv + \bar{v}_t^2 \int_0^{v_{max}} f_t(v)\,dv.$$

Further, from Eqs.(I.30) and (I.29a)

$$\bar{\sigma}_t^2 = \int_0^{v_{max}} v^2 f_t(v)\,dv - 2\bar{v}_t\bar{v}_t + \bar{v}_t^2 = \int_0^{v_{max}} v^2 f_t(v)\,dv - \bar{v}_t^2.$$

Restating this last result, we find that

$$\int_0^{v_{max}} v^2 f_t(v)\,dv = \bar{\sigma}_t^2 + \bar{v}_t^2. \tag{I.34}$$

Substituting Eq.(I.34) for the integral in Eq.(I.33), the desired relationship can now be written:

$$\bar{v}_x = \bar{v}_t + \frac{\bar{\sigma}_t^2}{\bar{v}_t}. \tag{I.35}$$

Furthermore, the variance $\bar{\sigma}_x^2$ can be expressed in terms of parameters of the time-dependent description. This relationship is given without derivation:

$$\bar{\sigma}_x^2 = \frac{\omega_t}{\bar{v}_t} - 2\bar{\sigma}_t^2 - \bar{v}_t^2 - \frac{\bar{\sigma}_t^4}{\bar{v}_t^2} \tag{I.36}$$

with

$$\omega_t = \int_0^{v_{max}} v^3 f_t(v)\,dv.$$

Example 18 (continued). Applying Eqs. (I.35) and (I.36) to Example 18, we obtain

$$\bar{v}_x = \frac{1}{2} v_{max} + \frac{\frac{1}{12} v_{max}^2}{\frac{1}{2} v_{max}} = \frac{2}{3} v_{max}$$

with

$$\omega_t = \int_0^{v_{max}} v^3 \frac{1}{v_{max}} \, dv = \frac{v_{max}^3}{4}$$

$$\bar{\sigma}_x^2 = \frac{1}{2} v_{max}^2 - \frac{1}{6} v_{max}^2 - \frac{1}{4} v_{max}^2 - \frac{1}{36} v_{max}^2 = \frac{1}{18} v_{max}^2$$

in correspondence with earlier results.

Example 20. Equation (I.35) also relates to the relationship between the route speed and the journey speed. The speed-time profile of a vehicle is described by the following equation:

$$v(t) = \begin{cases} 5 \text{ m/s}, & 0 \text{ s} \leq t \leq 10 \text{ s} \\ 5 \text{ m/s} + 2 \text{ m/s}^2 [t - 10 \text{ s}], & 10 \text{ s} < t \leq 30 \text{ s} \end{cases}$$

$$v_{max} = 5 + 2(30 - 10) = 45 \text{ m/s}.$$

The speed-time profile

$$v(t) = \begin{cases} v_0, & t_0 \leq t \leq t_1 \\ v_0 + b(t - t_1), & t_1 < t \leq 2 \end{cases}$$

corresponds to the speed-distance profile

$$v(x) = \begin{cases} v_0, & x_0 \leq x \leq x_1 \\ \sqrt{v_0^2 + 2b(x - x_1)}, & x_1 < x \leq x_2. \end{cases}$$

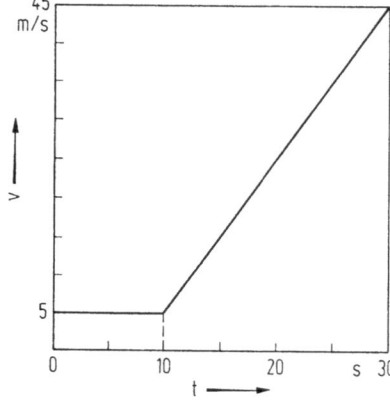

Fig. I.19

The distances x_0, x_1 and x_2 are obtained from the speed-time profile

$$x_0 = 0\,\text{m}$$

$$x_1 = x_0 + v_0 \cdot t_1 = 0\,\text{m} + 5\,\text{m/s} \cdot 10\,\text{s} = 50\,\text{m}$$

$$x_2 = x_1 + v_0(t_2 - t_1) + \frac{1}{2}b(t_2 - t_1)^2$$

$$= 50 + 5(30 - 10) + \frac{1}{2} \cdot 2(30 - 10)^2 = 550\,\text{m}.$$

After substitution we obtain the speed-distance profile

$$v(x) = \begin{cases} 5\,\text{m/s}, & 0 \leq x \leq 50\,\text{m} \\ \sqrt{25 + 4(x - 50)}\,\text{m/s}, & 50\,\text{m} < x < 550\,\text{m}. \end{cases}$$

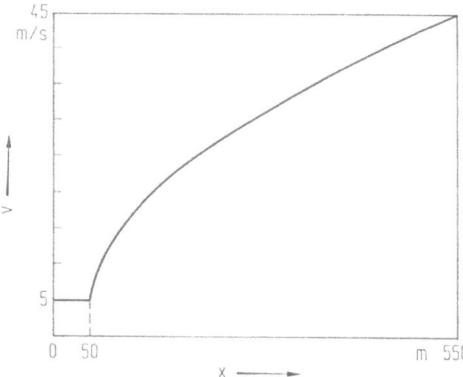

Fig. I.20

Mean and variance for the speed-time profile and for the speed-distance profile are obtained from Eqs. (I.22), (I.23), (I.24), and (I.25).

$$\hat{v}_t = \frac{1}{T}\left[\int_0^{t_1} v_0\,dt + \int_{t_1}^{t_2}(v_0 + b(t - t_1))\,dt\right]$$

$$= \frac{1}{T}\left(v_0 t\big|_0^{t_1} + \left[(v_0 - bt_1)t + \frac{1}{2}bt^2\right]\big|_{t_1}^{t_2}\right)$$

$$= \frac{1}{30}\left[5(10 - 0) + (5 - 2 \cdot 10)(30 - 10) + \frac{1}{2} \cdot 2(30^2 - 10^2)\right]$$

$$= 18.33\,\text{m/s}$$

$$\sigma_t^2 = \frac{1}{T} \left[\int_0^{t_1} v_0^2 dt + \int_{t_1}^{t_2} (v_0 + b(t-t_1))^2 dt \right] - \hat{v}_t^2$$

$$= -\hat{v}_t^2 + \frac{1}{T} \left[\int_0^{t_1} v_0^2 dt + \int_{t_1}^{t_2} ((v_0 - bt_1)^2 + 2bt(v_0 - bt_1) + b^2 t^2) dt \right]$$

$$= -\hat{v}_t^2 + \frac{1}{T} \left[v_0^2 t|_0^{t_1} + (v_0 - bt_1)^2 t|_{t_1}^{t_2} + b(v_0 - bt_1) t^2|_{t_1}^{t_2} + \frac{1}{3} b^2 t^3|_{t_1}^{t_2} \right]$$

$$= -18.33^2 + \frac{1}{30} [25(10-0) + (5-20)^2 (30-10)$$

$$+ 2(5-20)(30^2 - 10^2) + \frac{4}{3}(30^3 - 10^3)]$$

$$= 177.8 \text{ m}^2/\text{s}^2$$

$$\hat{v}_x = \frac{1}{X} \left\{ \int_0^{x_1} v_0 dx + \int_{x_1}^{x_2} [v_0^2 + 2b(x-x_1)]^{1/2} dx \right\}$$

$$\hat{v}_x = \frac{1}{X} \left\{ v_0 x \Big|_0^{x_1} + \left[\frac{1}{2b\left(\frac{1}{2}+1\right)} (2bx + (v_0^2 - 2bx_1)) \right]^{3/2} \Big|_{x_1}^{x_2} \right\}$$

$$= \frac{1}{550} \left\{ 5(50-0) + \frac{1}{4 \cdot \frac{3}{2}} [4 \cdot 550 + 25 - 200]^{3/2} - \frac{1}{4 \cdot \frac{3}{2}} [4 \cdot 50 + 25 - 200]^{3/2} \right\}$$

$$= 28 \text{ m/s},$$

$$\sigma_x^2 = \left\{ \frac{1}{X} \left[\int_0^{x_1} v_0^2 dx + \int_{x_1}^{x_2} [v_0^2 + 2b(x-x_1)] dx \right] \right\} - \hat{v}_x^2$$

$$= -\hat{v}_x^2 + \frac{1}{X} [v_0^2 x|_0^{x_1} + (v_0^2 - 2bx_1) x|_{x_1}^{x_2} + bx^2|_{x_1}^{x_2}]$$

$$= -28^2 + \frac{1}{550} [25(50-0) + (25-200)(550-50) + 2(550^2 - 50^2)]$$

$$= 150 \text{ m}^2/\text{s}^2.$$

The results can be checked by means of the relationship between route speed and journey speed given in Eq. (I.35).

$$\bar{v}_x = \bar{v}_t + \frac{\bar{\sigma}_t^2}{\bar{v}_t}$$

$$\bar{v}_x = 18.33 + \frac{177.8}{18.33} = 28.0 = \hat{v}_x.$$

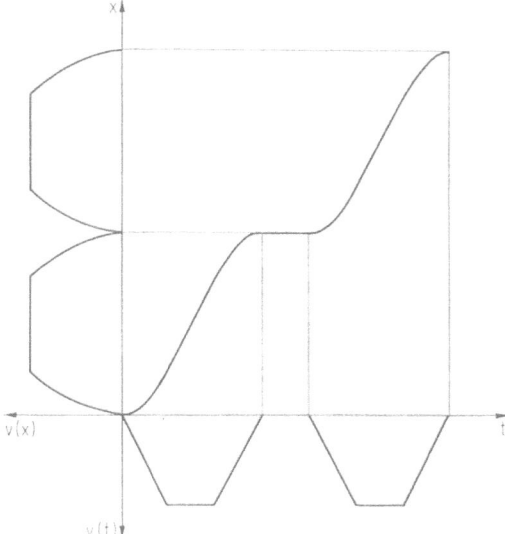

Fig. I.21

For a given trajectory, there is a difference between the mean values and the variances depending on whether the motion is analysed as a function of time or of distance. This difference is clarified if one considers a journey having an intermediate stop (Fig. I.21): While the duration of this intermediate stop obviously enters the calculation of the journey speed, this is not the case for the calculation of the route speed. The value of the route speed is independent of how long a vehicle stays at some particular point in its route with speed $v = 0$. The route speed can be useful in those situations where one wants to eliminate the effects of intermediate stops (e.g. waiting times at traffic signals).

Clearly, slowness w can also be represented as a function of distance or as a function of time; the respective mean values (with $t_1 - t_0 = T$ and $x_1 - x_0 = X$) are calculated as

$$\hat{w}_x = \frac{1}{X} \int_{x_0}^{x_1} w(x)\,dx = \frac{\int_{x_0}^{x_1} \frac{dt}{dx}\,dx}{\int_{x_0}^{x_1} dx} = \frac{\int_{t_0}^{t_1} dt}{\int_{x_0}^{x_1} dx} = \frac{T}{X} = \frac{1}{\hat{v}_t}$$

and

$$\hat{w}_t = \frac{1}{T} \int_{t_0}^{t_1} w(t)\,dt.$$

All four mean values of speed and slowness can be summarized:

$$\hat{v}_t = \frac{\int\limits^{T} v(t)\,dt}{\int\limits^{X} w(x)\,dx}; \quad \hat{w}_t = \frac{\int\limits^{T} w(t)\,dt}{\int\limits^{X} w(x)\,dx}$$

$$\hat{w}_x = \frac{\int\limits^{X} w(x)\,dx}{\int\limits^{T} v(t)\,dt}; \quad \hat{v}_x = \frac{\int\limits^{X} v(x)\,dx}{\int\limits^{T} v(t)\,dt}.$$

Thus, there is reciprocity not only between the time-based speed and the distance-based slowness, but also between the mean value of speed over time (the journey speed) and the mean value of slowness over distance. (For a similar example of this kind of reciprocity, see Sect. II.2.3.2).

Therefore, in order to calculate the four mean values it suffices to know the expressions for four definite integrals.

> **Example 21.** Consider again the motion of a vehicle which accelerates with constant acceleration from $v = 0$ to $v = v_{max}$. Starting from $t_0 = 0$ and $x_0 = 0$.
>
> $$v(t) = bt; \qquad\qquad bT = v_{max}$$
>
> $$x(t) = \frac{1}{2}bt^2 = \frac{v^2}{2b}; \qquad t(x) = \sqrt{\frac{2x}{b}}$$
>
> $$v(x) = \sqrt{2bx}; \qquad\qquad \sqrt{2bX} = v_{max}$$
>
> $$w(t) = \frac{1}{v(t)} = \frac{1}{bt}$$
>
> $$w(x) = \frac{1}{v(x)} = \frac{1}{\sqrt{2bx}}.$$
>
> From this we obtain the desired integrals:
>
> $$\int_0^T v(t)\,dt = \int bt\,dt = \frac{1}{2}bT^2 = \frac{v_{max}T}{2} \qquad \left(= \int_0^X dx = X\right)$$
>
> $$\int_0^X v(x)\,dx = \int \sqrt{2bx}\,dx = \frac{2}{3}Xv_{max}$$
>
> $$\int_0^X w(x)\,dx = \int \frac{dx}{\sqrt{2bx}} = \sqrt{\frac{2X}{b}} \qquad \left(= \int_0^T dt = T\right)$$
>
> $$\int_0^T w(t)\,dt = \int \frac{dt}{bt} = \frac{1}{b}(\ln T + \infty) = \infty.$$
>
> Then we have
>
> $$\hat{v}_t = \frac{v_{max}T}{2T} = \frac{1}{2}v_{max}; \quad \hat{w}_x = \frac{2T}{v_{max}T} = \frac{2}{v_{max}}$$
>
> $$\hat{v}_x = \frac{2Xv_{max}}{3X} = \frac{2}{3}v_{max}; \quad \hat{w}_t = \frac{\infty}{T} = \infty.$$

I.2.3 The Distribution of Acceleration

In general, not only vehicle speeds but also the other related kinematic parameters have statistical distributions. Of particular importance is the distribution of the

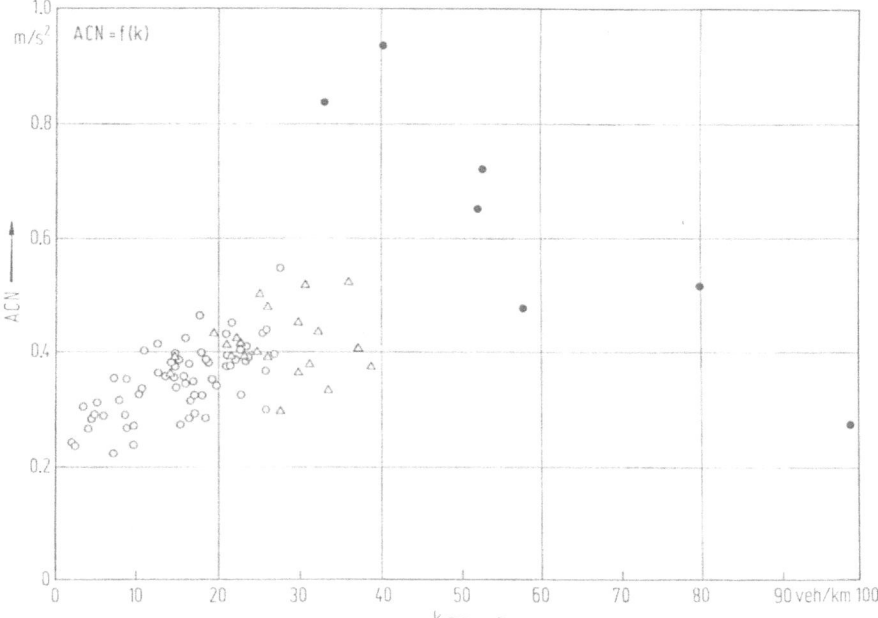

Fig. I.22. (From [67])
Fig. I.22. (From [67])

acceleration values during a journey. Let T be the total journey time of a vehicle. The mean acceleration during the journey is:

$$\bar{b} = \frac{1}{T} \int_0^T b(t) \, dt$$

and the corresponding standard deviation, sometimes referred to as acceleration noise (ACN), is

$$\text{ACN} = \sqrt{\frac{1}{T} \int_0^T [b(t) - \bar{b}]^2 \, dt}. \tag{I.37}$$

In the same way, the parameters of the distribution of acceleration can be specified as functions of distance.

Acceleration noise is a good indicator for the smoothness of the traffic flow. Measurements show that acceleration noise depends on traffic density (see Fig. I.22); starting from a minimum level when $k \to 0$, it increases with increasing traffic density to attain a maximum in the same region as flow in the fundamental diagram reaches a maximum (see also Sect. II.2.6.3), after which it decreases.

Chapter II. The Motion of Several Vehicles on a Road

When several vehicles are moving on a roadway, we will speak, in what follows, of a traffic stream. The existence and influence of intersections and other stopping points of any sort will be ignored. Therefore, traffic flow will be considered on road sections without stopping points, hereafter referred to as links. The lines of motion of single elements (vehicles) of the traffic stream can intersect if overtaking occurs (Fig. II.1).

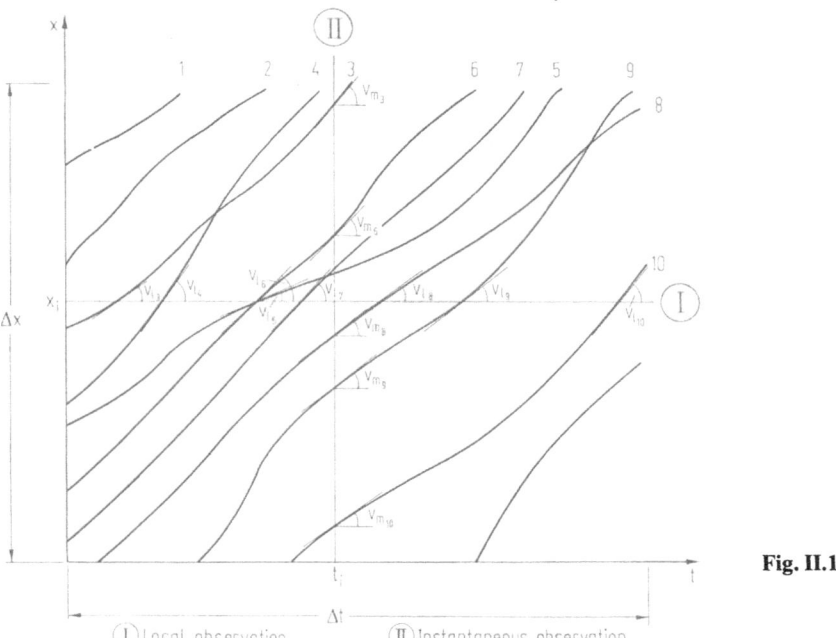

Fig. II.1

In this context we label the measurements of a traffic parameter at a measuring point fixed in distance but over a finite time interval, as a "local observation", and a measurement at a fixed point in time but over a finite distance as an "instantaneous observation".

Deterministic models for describing a traffic stream presume the knowledge of the motion of every single vehicle. With these models one can specify the parameters of the traffic stream at every location and at every point in time. With certain transport systems, as for example, roads or inland waterways one can explain, at least in theory, why a vehicle is found at a specified place at a specified time. But the numbers of factors influencing motion are so large, that for an external observer the motion appears to be random. Thus, for practical purposes, traffic flow becomes a stochastic process. Statements concerning certain characteristic properties (parameters) are only possible as probabilities. Therefore, before passing over to the definition and description of such parameters, probability distributions are discussed.

II.1 Distributions and Their Parameters

In Sect. I.2 the motion of a single vehicle was described in terms of the means and variances of the empirical density functions. These in turn are derived from speed-time and speed-distance profiles. Under certain assumptions[1], the relative frequency can be interpreted as a probability though, for the examples in Sect. I.2 these assumptions are in general not true.

Let X be any random variable, which is allowed to assume only discrete values. Then the probability that X has a specific value x_i is

$$P[X = x_i] = P_i \tag{II.1}$$

and the probability density function of the discrete random variable is defined as

$$P(x) = \begin{cases} P[X = x] = P_i, & \text{for } x = x_i \\ 0, & \text{otherwise.} \end{cases} \tag{II.2}$$

The distribution function of X is defined as the probability that X takes any value less than equal to x is

$$P[X \leq x] = F(x) = \sum_{x_i \leq x} P_i. \tag{II.3}$$

In particular

$$P[X \leq +\infty] = \sum_{x_i \leq +\infty} P_i = 1. \tag{II.4}$$

The term $dF(x)$ denotes the probability density function: thus the probability that the random variable X takes on a value x_i is $dF(x_i)$. That is

$$dF(x) = f(x); \quad dF(x_i) = P[X = x_i] = p_i.$$

The expected value of X is, by definition,

$$E(X) = \sum_{x_i \leq +\infty} x_i p_i. \tag{II.5}$$

---

1 One assumption is that the kinds of variables under consideration are random, i.e. they assume unpredictable values with each repetition of the underlying experiments. This of course is not the case for deterministic variables (as in the examples in Sect. I.2). One further assumption is that in contrast to the preceding examples which deal with only a single sample, the values of all of the relevant parameters of the population (i.e. the sampling universe) are known.

As already encountered in Example 17, the arithmetic mean of a sample of size n is defined as

$$\bar{x} = \frac{1}{n} \sum_{i=1}^{n} x_i.$$

This expectation operation defined in Eq. (II.5) gives the population mean, which is a constant. In contrast, the arithmetic mean, $\bar{x}$, for a sample, which is drawn from the underlying population is a random variable and is only an estimate of the population mean, $E(X)$. The population variance is defined as

$$\sigma^2 = \sum_{x_i \leq +\infty} [x_i - E(X)]^2 p_i. \tag{II.6}$$

Again as already encountered in Example 17, the sample variance is defined as

$$s^2 = \frac{1}{n-1} \sum_{i=1}^{n} (x_i - \bar{x})^2.$$

Any particular value of s^2 is only an estimate of the population variance σ^2. Again it is emphasized that σ^2 is a fixed and constant parameter of the population, whereas s^2 is a random variable whose value depends on the sample selected.

Example 22. One discrete distribution frequently used in traffic science is the Poisson distribution (see Sect. II.2.4 for an application of this distribution). The Poisson distribution is used for example to calculate the probability that, under "stationary" flow conditions (see Sect. II.2.1) m vehicles pass a stationary observer in a time period Δt. The Poisson probability mass function is

$$P[M = m] = \frac{(\lambda \Delta t)^m}{m!} e^{-\lambda \Delta t}$$

and the distribution function is

$$P[M \leq m] = \sum_{m_i \leq m} \frac{(\lambda \Delta t)^{m_i}}{m_i!} e^{-\lambda \Delta t}.$$

The quantity $\lambda \Delta t$ is the average number of vehicles which would be counted during a fixed time period of Δt (for the definition of λ, see Sect. II.2.1). The shape of the probability mass function depends on the size of the quantity $\lambda \cdot \Delta t$ (Fig. II.2):

For $\lambda = 0.5$ veh/s, the probability that $m = 3$ veh. are counted in a time interval $\Delta t = 10$ s is

$$P[M = 3] = \frac{(0.5 \cdot 10)^3}{3!} e^{-0.5 \cdot 10} = 0.1404$$

and the probability, that less than four vehicles will be counted is

$$P[M \leq 3] = \sum_{m_i = 0}^{m_i = 3} \frac{(0.5 \cdot 10)^{m_i}}{m_i!} e^{-0.5 \cdot 10} = 0.2650.$$

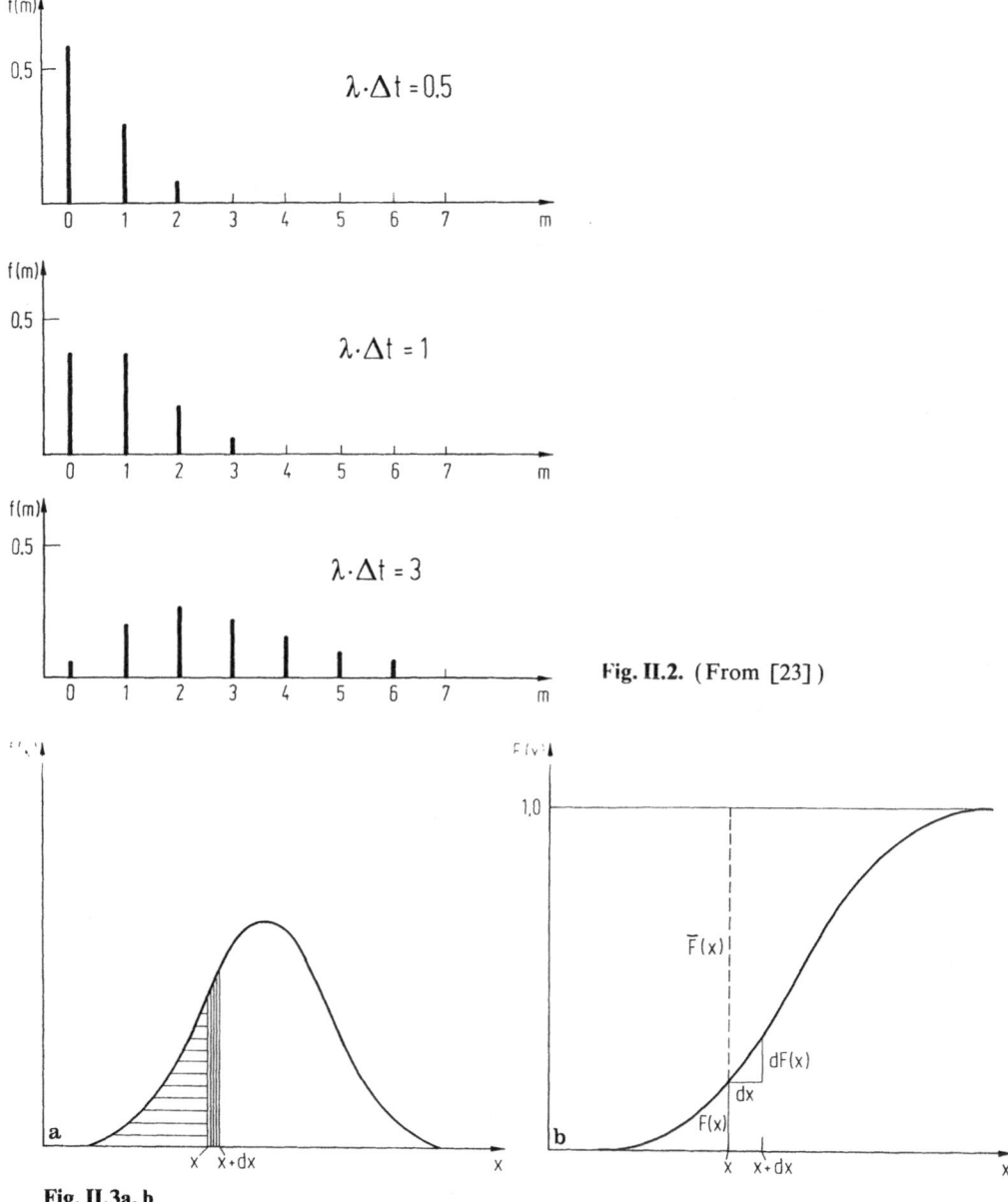

Fig. II.2. (From [23])

Fig. II.3a, b

If a random variable X is not restricted to discrete values, but can assume any value in some interval, then X is a continuous random variable. The probability that X takes on a value between x and x + dx (see Fig. II.3a) is

$$P[x \leqq X \leqq x + dx] = \int_{x}^{x+dx} f(y)\,dy. \tag{II.7}$$

The term f(x) is defined to be the probability density function of the random variable X.

The probability that X assumes a value less than or equal to x (Fig. II.3b)

$$P[X \leq x] = F(x) = \int_{-\infty}^{x} f(y) \, dy \tag{II.8}$$

is defined by analogy to discrete random variables as the cumulative distribution function of the random variable X, or more simply, as the distribution function. The probability that X takes a value larger than x (see Fig. II.3b) is

$$P[X > x] = \bar{F}(x) = \int_{x}^{\infty} f(y) \, dy \tag{II.9}$$

and is referred to as the complementary distribution function.

In particular,

$$P[X \leq +\infty] = \int_{-\infty}^{+\infty} f(y) \, dy = F(x) + \bar{F}(x) = 1. \tag{II.10}$$

The probability density function $f(x)$ is therefore the derivative of the distribution function $F(x)$: $f(x) = dF(x)/dx$. The probability that the random variable X has a value between x and $x + dx$ can therefore also be expressed as

$$dF(x) = f(x) \, dx.$$

The expected value of a continuous random variable X is defined as

$$E(X) = \int_{-\infty}^{+\infty} x f(x) \, dx. \tag{II.11}$$

As in the case of discrete random variables, the arithmetic mean of a random sample is an estimate of the population mean $E(X)$.

The calculation of the expected value is analogous to the determination of the x-coordinate of the center of gravity of the area under the probability density function. The sum of the moments of the elemental areas $f(x) dx$

$$\int_{-\infty}^{+\infty} x f(x) \, dx$$

is set equal to the moment of the entire area about the ordinate

$$\bar{x} \int_{-\infty}^{+\infty} f(x) \, dx.$$

Substituting Eq. (II.10) in the above expression results in Eq. (II.11).

The shape of the probability density function is determined by the distribution of the random variable X about its mean. One approach is to describe the shape of the distribution using such quantities as

$$\int_{-\infty}^{+\infty} [x - E(X)]^n f(x) \, dx. \tag{II.12}$$

This integral is defined as the n-th central moment of a continuous distribution. In particular we have for $n = 0$

$$\int_{-\infty}^{+\infty} [x - E(X)]^0 f(x) \, dx = \int_{-\infty}^{+\infty} f(x) \, dx = 1$$

that the 0-th central moment is equal to the area under the probability density function;

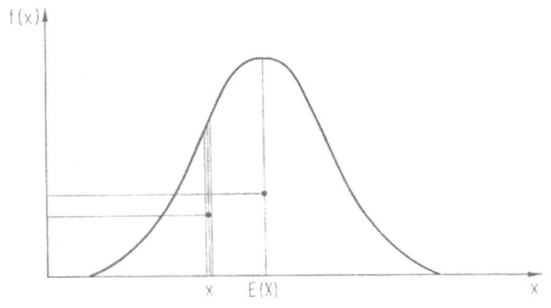

Fig. II.4

for $n=1$:

$$\int_{-\infty}^{+\infty} [x-E(X)]^1 f(x)dx = \int_{-\infty}^{+\infty} xf(x)dx - E(X)\int_{-\infty}^{+\infty} f(x)dx$$

$$= E(X) - E(X) = 0$$

for $n=2$:

$$\sigma^2 = \int_{-\infty}^{+\infty} [x-E(X)]^2 f(x)dx$$

$$= \int_{-\infty}^{+\infty} [x^2 - 2xE(X) + E(x)^2]f(x)dx.$$

Generalizing Eq. (II.11) to the n-th moment of a random variable X we define

$$\int_{-\infty}^{+\infty} x^n f(x)dx = E(X^n)$$

and go on to use this result to express the variance as

$$\sigma^2 = \int_{-\infty}^{+\infty} x^2 f(x)dx - 2E(X)\int_{-\infty}^{+\infty} xf(x)dx + E(X)^2\int_{-\infty}^{+\infty} f(x)dx$$

$$= E(X^2) - 2[E(X)]^2 + [E(X)]^2 = E(X^2) - [E(X)]^2. \tag{II.13}$$

An estimate for the population variance σ^2 can be obtained as in the case of discrete random variables by computing the sample variance.

Figure II.5 illustrates the relationship between the shape of the probability density function and its variance, keeping the mean constant:

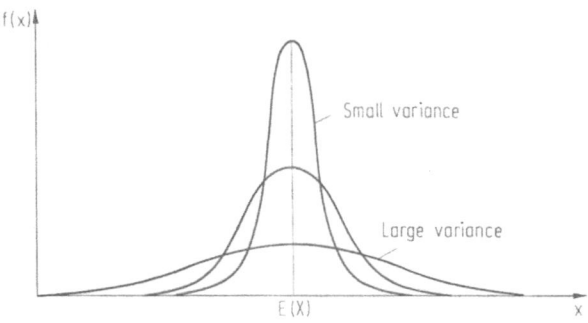

Fig. II.5

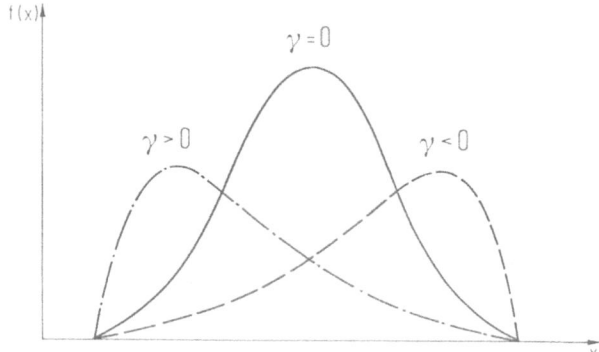

Fig. II.6

Finally, the skewness (or asymmetry) is defined in terms of the third central moment (Fig. II.6) as

$$\gamma = \frac{1}{\sigma^3} E[(X - E(X))^3]. \tag{II.14}$$

For a symmetric probability density function

$$E[(X - E(X))^3] = 0$$

so that

$$\gamma = 0.$$

A probability density function is negatively skewed when $\gamma < 0$ and is positively skewed when $\gamma > 0$.

Example 23. With the help of speed measuring apparatus, the speeds of 108 vehicles at a point on an urban road are observed.

Speed classes (km/h)	Frequencies	Relative frequencies $f_t(v_i)$	Cumulative relative frequencies $F_t(v_i)$
20.00–24.99	0	0.00	0.00
25.00–29.99	1	0.01	0.01
30.00–34.99	2	0.02	0.03
35.00–39.99	14	0.13	0.16
40.00–44.99	15	0.14	0.30
45.00–49.99	33	0.31	0.61
50.00–54.99	26	0.24	0.85
55.00–59.99	5	0.05	0.89
60.00–64.99	7	0.06	0.95
65.00–69.99	1	0.01	0.06
70.00–74.99	1	0.01	0.97
75.00–79.99	3	0.03	1.00
80.00–84.99	0	0.00	1.00

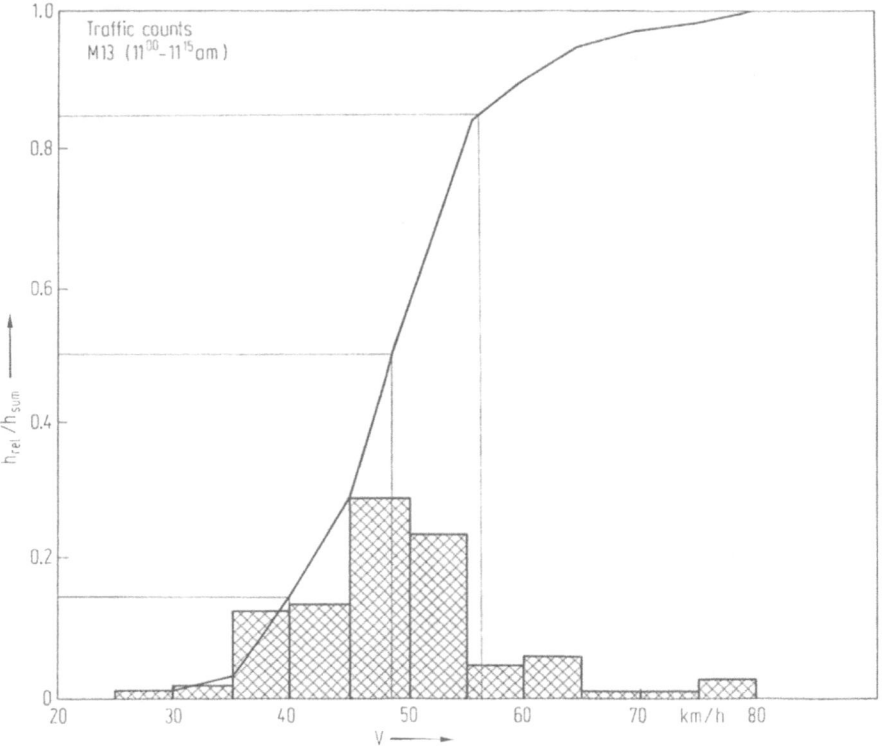

Fig. II.7

The mean speed is 49 km/h and the standard deviation 9 km/h. The 15-th, 50-th and 85-th percentiles have the values 40 km/h, 48 km/h and 56 km/h, respectively.

The observations may be represented as a relative frequency histogram and the corresponding cumulative distribution curve (Fig. II.7). For the calculation of mean and variance, see Example 17.

II.2 Parameters for Describing a Traffic Stream

II.2.1 Discrete Flows and Local Measurements

A traffic stream is observed from a fixed measuring point x_i during some time interval Δt (Fig. II.1) and a continuous record is kept (e.g. using a time recorder) of the times when vehicles pass. Let us define the stream function $\Phi_{x_i}(t)$ as the accumulated vehicle count at observation point x_i during the time interval $(0,t)$. $\Phi_{x_i}(t)$ can only increase by integer increments. Such a function represents a non-decreasing, integer valued stochastic process, or a discrete flow process, and in the following is called CUSUM function.

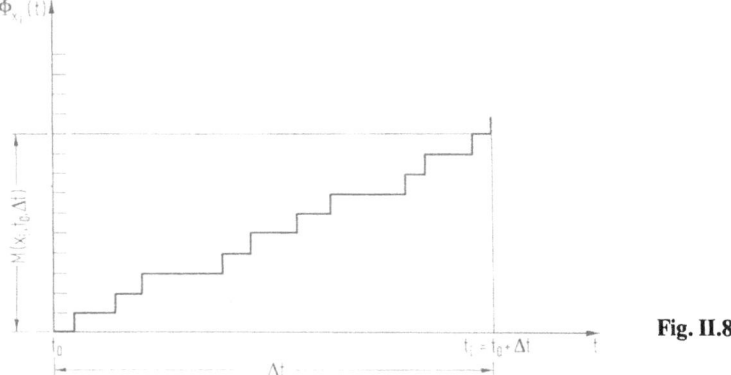

Fig. II.8

The quotient

$$q = \frac{\Phi_{x_i}(t_i) - \Phi_{x_i}(t_0)}{\Delta t} = \frac{M(x_i, t_0, \Delta t)}{\Delta t} \qquad (II.15)$$

[veh/time-interval] is defined as the traffic volume. The statement of traffic volume is not complete without a statement of the time interval over which the vehicle count M was taken. In general, one should not interpolate to shorter time intervals or extrapolate to longer time intervals.

The intensity of the traffic stream at a point x_i is defined as the limit:

$$\lim_{\Delta t \to 0} \frac{P[M(x_i, t, \Delta t) \geq 1]}{\Delta t} = \lambda_{x_i}(t) \qquad [\text{veh/unit-time}]. \qquad (II.16)$$

The intensity is in general a function of time; when this is the case, the process is said to be non-stationary. When the intensity is time-independent the process is said to be stationary. Whether or not stationarity exists may be determined by statistical test procedures. The quantity $\lambda_{x_i}(t)\Delta t$ can be interpreted as the probability that a vehicle passes the observation point x_i in some arbitrarily small time interval Δt. As long as the flow rate is not too high the probability of more than one vehicle passing $P[M(x_i, t, t) > 1]$ vanishes as Δt approaches 0. I.e. two vehicles do not pass the point x_i at the same time, even when they are travelling in adjacent lanes. Therefore,

$$\lambda_{x_i}(t) = \lim_{\Delta t \to 0} \frac{P[M(x_i, t, \Delta t) = 1]}{\Delta t}.$$

If the process is stationary and if λ_{x_i} is a known constant, then the expected number of vehicles in an interval Δt is

$$E(M) = \lambda \Delta t \qquad [\text{veh}]. \qquad (II.17)$$

Example 24. Let $\lambda = 0.2$ veh/s. Then, in a 10 min period, the expected value of M is $E(M) = 0.2 \cdot 600$ or 120 veh and the volume is thus $q = 120$ veh/10 min.

When λ is estimated from a local measurement (see Sect. II.2.3.2), the value of M can be estimated for time intervals smaller or larger than the observation period

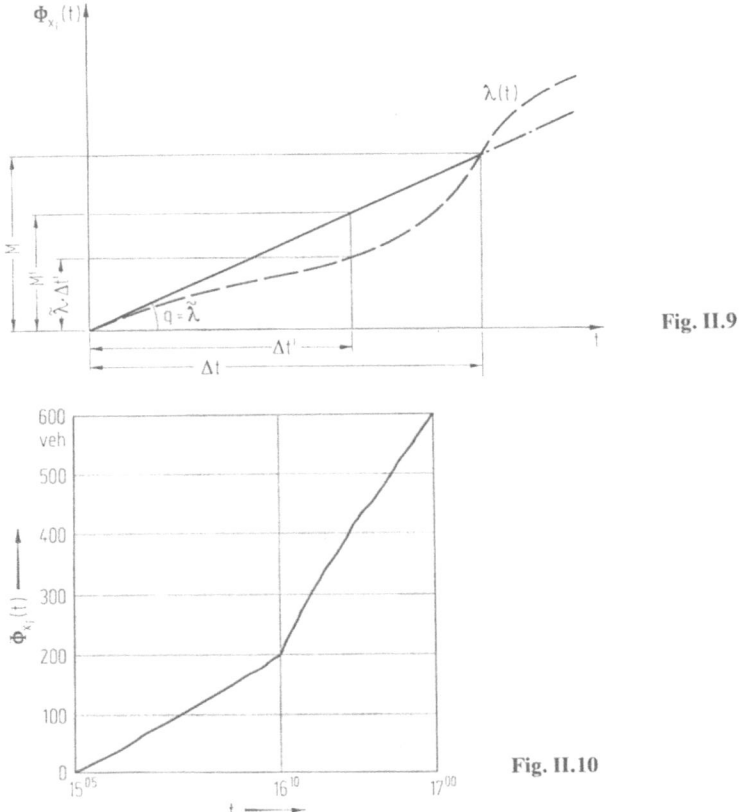

Fig. II.9

Fig. II.10

only when the process is stationary (see Sect. II.9). When the process is not stationary, M' differs from $\tilde{\lambda}\Delta t'$ by more than random fluctuations, where $\tilde{\lambda}$ is the average intensity during time period Δt.

Example 25. 300 motor vehicles are counted on a motorway during a 5 min period; $q = 300$ veh/5 min. Only in the case of stationarity can the volume be extrapolated as being 3 600 veh/h.

Whether or not a traffic stream is stationary can be determined from the CUSUM function $\Phi_{x_i}(t)$ (see Fig. II. 10): Stationarity can be assumed for those portions of $\Phi_{x_i}(t)$ which are not significantly different from a straight line of the form $\lambda_k t + a_k$. This hypothesis can be tested by applying the Wilcoxon-test.

Imagine that the traffic stream is observed not just at one point x_i, but at each point within some distance interval as a function of time and distance. This is denoted by $n(x,t)$; $\Phi_x(t)$ is obtained from this as a cross-section parallel to the t-axis. One can picture $n(x,t)$ as looking something like a set of steps, each edge represents the trajectory $x = f(t)$ of a single vehicle and each step represents the occurence of a vehicle (see Fig. II.11).

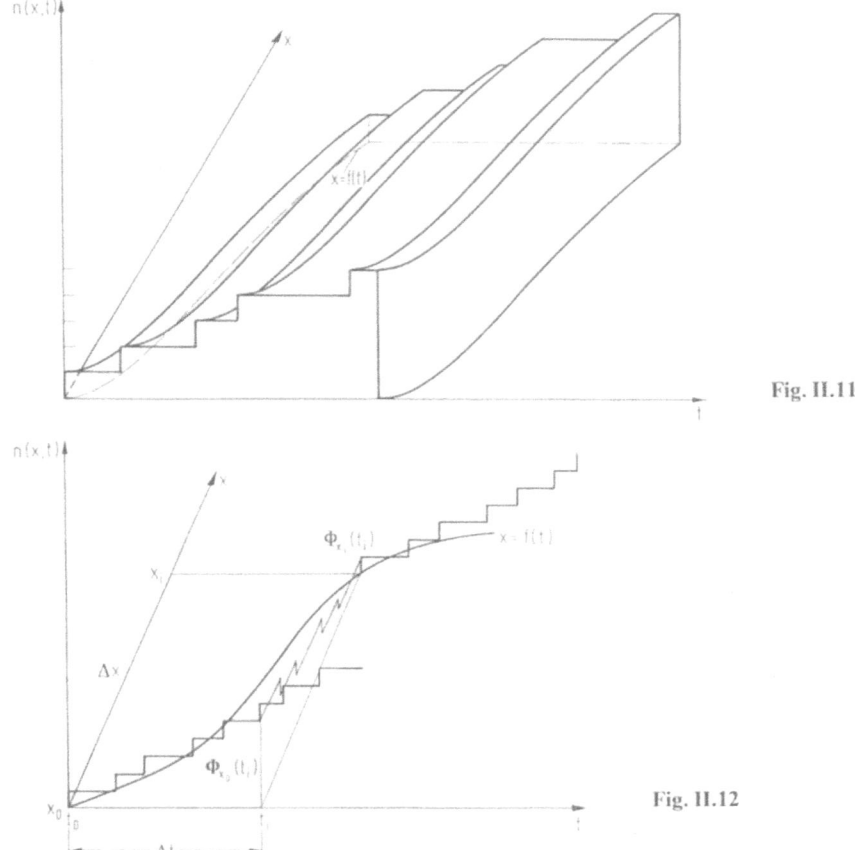

Fig. II.11

Fig. II.12

From the representation it is clear that if one looks at $\Phi_{x_0}(t)$ and at Φ_{x_1} (see Fig. II.12)[1], the number of vehicles which, at time t_i, are found in the interval $\Delta x = x_i - x_0$ is

$$N(t_i, x_0, \Delta x) = \Phi_{x_0}(t_i) - \Phi_{x_i}(t_i)$$

(see Fig. II.13), and, further, we define

$$k = -\frac{\Phi_{x_i}(t_i) - \Phi_{x_0}(t_i)}{\Delta x} = \frac{N(t_i, x_0, \Delta x)}{\Delta x}. \qquad (II.18)$$

---

1 One must be careful that the counting begins at each counting location with the passage of the same vehicle (assuming no overtaking). If counting begins at all locations at the same time, e.g. at $t = t_0$, one must account for the vehicles which pass, for example, the second counter, during the time interval Δt

$$\Delta t = \int_{x_0}^{x_i} \frac{dx}{v(x)} = \int_{x_0}^{x_i} w(x) dx,$$

during which the first vehicle counted at x_0, traverses the distance $x = x_i - x_0$. The problem is that the function $\Phi_{x_i}(t)$ includes these vehicles which are counted during the interval Δt, but $\Phi_{x_0}(t)$ does not (see Fig. II.14).

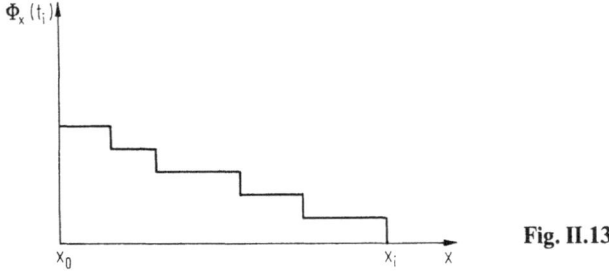

Fig. II.13

The quotient k is called the "traffic density" and must be accompanied by a statement of the distance Δx in which the vehicle count N was made.

The concentration at time t_i is defined as the limit

$$\lim_{\Delta x \to 0} \frac{P[N(t_i,x,\Delta x) \geq 1]}{\Delta x} = \varkappa_{t_i}(x). \tag{II.19}$$

Depending of whether the concentration is a function of distance or is independent of distance, the process is defined as being non-stationary or stationary in distance.

The quantity $\varkappa_{t_i}(x)\,dx$ can be interpreted as the probability that, in any arbitrarily small distance interval, one or more vehicles will be present at time t_i.

As x approaches zero the probability $P[N(t_i,x,\Delta x) > 1]$ vanishes, so that Eq. (II.19) becomes

$$\varkappa_{t_i}(x) = \lim_{\Delta x \to 0} \frac{P[N(t_i,x,\Delta x) = 1]}{\Delta x}.$$

Further relationships between $\varkappa$ and k follow the same pattern as for λ and q. While M (and thus q) can be calculated as the difference of the ordinates of a staircase function $\Phi_{x_i}(t)$ which is obtained by cutting the surface n(x,t) parallel to the t-axis, N (and thus k) is calculated as the difference of the ordinates of two staircase functions $\Phi_{x_0}(t)$ and $\Phi_{x_i}(t)$, at the same time point t_i, from a cut through n(x,t) parallel to the x-axis.

II.2.2 Discrete Flows and Instantaneous Measurements

The time-distance curves can be summed not only over the time axis (as in Fig. II.14) but also over the distance axis (as in Fig. II.16). One then obtains a function n'(x,t) as a reflection. An instantaneous observation now corresponds to a cross-section through the distance function n'(x,t) parallel to the x-axis.

A traffic stream is observed over a distance Δx (e.g. by aerial photography) at some time point t_i, and the positions of all vehicles found in Δx are recorded (Fig. II.1). Let us define the stream function $\Psi_{t_i}(x)$ as the accumulated vehicle count at time t_i for the interval (0,x), where the x-axis corresponds to the direction of motion. At each point at which a vehicle is located, the function $\Psi_{t_i}(x)$ increases by one unit (analogously to Fig. II.15). Such a function, as in

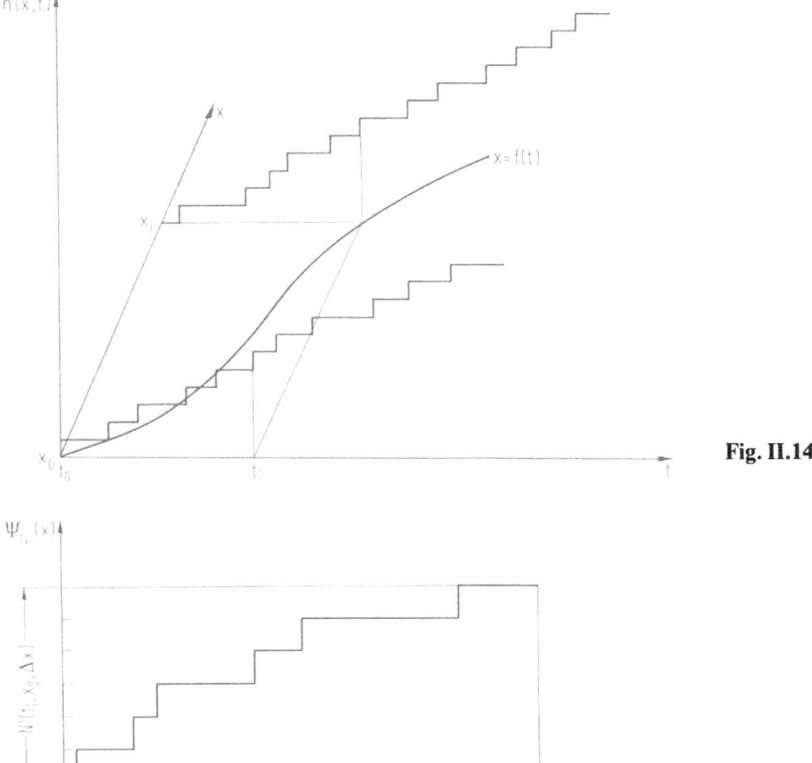

Fig. II.14

Fig. II.15

Sect. II.1, is a non-decreasing, integer valued stochastic process, or a discrete flow process.

The traffic density is obtained, in analogy with Eq. (II.15) as a difference in the ordinates of the CUSUM function $\Psi_{t_i}(x)$ at time t_i:

$$k = \frac{\Psi_{t_i}(x_i) - \Psi_{t_i}(x_0)}{\Delta x} = \frac{N'(t_i, x_0, \Delta x)}{\Delta x} \quad [\text{veh/distance}]. \tag{II.20}$$

From this concentration at time t_i can be computed as in Sect. II.2.1.

As in Sect. II.2.1, the number M' of vehicles passing the point x_i during the time interval $\Delta t = t_i - t_0$ can be computed, in this case from two instantaneous observations $\Psi_{t_0}(x)$ and $\Psi_{t_i}(x)$ (see Fig. II.16).

From

$$M'(x_i, t_0, \Delta t) = \Psi_{t_0}(x_i) - \Psi_{t_i}(x_i)$$

one can compute

$$q = -\frac{\Psi_{t_i}(x_i) - \Psi_{t_0}(x_i)}{\Delta t} = \frac{M'(x_i, t_0, \Delta t)}{\Delta t}. \tag{II.21}$$

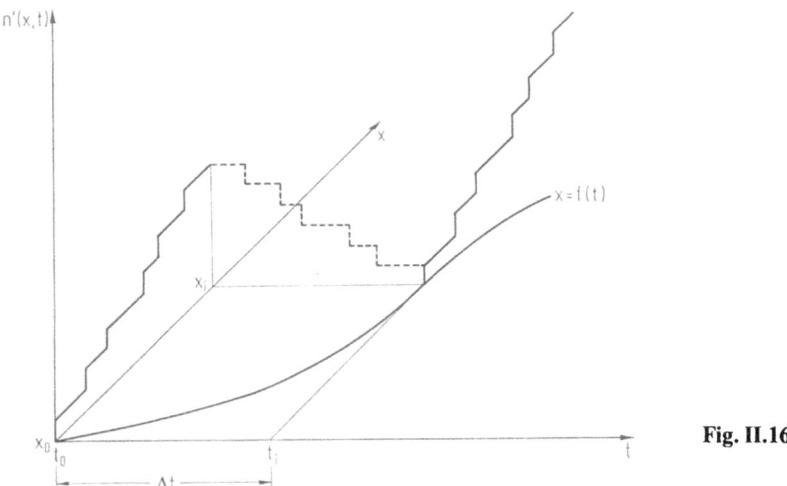

Fig. II.16

The difference between Eqs. (II.21) and (II.15) should be carefully noted. Let us compare the parameters obtained from the two measurement methods. For the functions $n(x,t)$, resp. $n'(x,t)$, we have the following possible configurations of observations:

Instantaneous measurements $[n'(x,t)]$	Spot measurements $[n(x,t)]$
Two measurements at t_0 and $t_i = t_0 + \Delta t$ $$q = -\frac{\Psi t_i(x_i) - \Psi t_0(x_i)}{\Delta t}$$	One measurement at x_i over Δt $$q = \frac{\Phi x_i(t_i) - \Phi x_i(t_0)}{\Delta t}$$
One measurement at t_i over Δx $$k = \frac{\Psi t_i(x_i) - \Psi t_i(x_0)}{\Delta x}$$	Two measurements at x_0 and $x_i = x_0 + \Delta x$ $$k = -\frac{\Phi x_i(t_i) - \Phi x_0(t_i)}{\Delta x}$$

Example 26. At two census points on a one-way carriageway separated by a distance of 5 km vehicles are counted, starting with the same vehicle. At a particular time (say at 11:45) 217 vehicles have passed census point 1 and 112 vehicles have passed census point 2. Then at 11:45 there are $217 - 112 = 105$ vehicles between the census points and the density is

$$\varkappa = \frac{105}{5} = 21 \text{ veh/km}.$$

(Assuming an even distribution of vehicles and the number of vehicles that reference vehicle has passed equals the number that have passed it.)

II.2.3 Speed Distributions

It was pointed out in Chap. I, that the speed of an individual vehicle is, in general, not constant over either time or distance. Thus, when one measures the speeds of a number of vehicles, the speeds are generally different, so that one obtains a distribution of speeds which can be described using the usual methods of mathematical statistics (see Sect. II.1).

No assumption will be made as to the shape of the distribution. This depends on the traffic mix, the road conditions, the measuring method (see Sect. II.2.5.2) etc.

II.2.3.1 The Instantaneous Speed Distribution

Measurements at some specific point in time are defined as instantaneous measurements (see Sect. II.1). The underlying speed distribution resulting from instantaneous measurements of speed will be denoted as $G_m(v)$, and the corresponding probability density function denoted as $g_m(v)$, where $g_m(v)dv = dG_m(v)$. For a discrete distribution we will use

$$\Delta G_m(v) = \begin{cases} P_m(v = v_i), & \text{for} \quad v = v_i \\ 0, & \text{otherwise} \end{cases}$$

(see Sect. II.1).

Truely instantaneous speed measurements are rarely feasible. One can imagine the situation where all vehicles carry a large speedometer on their roofs: an aerial photo would then provide such a measurement. The arithmetic mean of a set of such speed measurements is

$$\bar{v}_m = \frac{1}{N} \sum_{i=1}^{N} v_i,$$

or, if there are n_i vehicles having speed v_i and a total of N vehicles,

$$\bar{v}_m = \sum_{i=1}^{k} v_i \Delta G_m(v_i).$$

This is called a space-mean speed. The sample mean v_m estimates the population mean

$$E_m(v) = \int_0^\infty v g_m(v) dv = \int_0^\infty v \, dG_m(v). \tag{II.22}$$

II.2.3.2 The Local Speed Distribution

Measurements at some fixed measuring point are defined as local measurements (see Sect. II.1). The underlying distribution function for local measurements of speed is denoted as $G_l(v)$. Good approximations to local speed measurements are possible, for example using radar, which shows the desired speed measurement directly. The arithmetic mean, as above, is

$$\bar{v}_l = \frac{1}{M} \sum_{i=1}^{M} v_i = \sum_{i=1}^{k} v_i dG_l(v_i) \tag{II.23}$$

and is an estimate of the population mean

$$E_1(v) = \int_0^\infty vg_1(v)dv = \int_0^\infty v\,dG_1(v),$$

$\bar{v}_1$ is called a time-mean speed. The difference between local and instantaneous measurements of speed is illustrated in the following example. It will be assumed that N vehicles travel on a circular road of length L. Their speed distribution is described by $G_m(v)$. Each vehicle travels at its desired speed, which remains constant. This assumes that each vehicle can at any time immediately carry out all necessary overtaking manoeuvers. This situation is defined as that of free flow.

The traffic density is $k = N/L$ [veh/distance]. The number of these vehicles with speed v is $dk(v) = kdG_m(v)$. Now consider the traffic volume q at a location x during a time interval T.

A vehicle travelling at speed v requires the time $t = L/v$ for one circuit. In this time period a stationary observer would see, just once, each vehicle having a speed in the differential interval, $(v,v+dv)$. Thus, a volume of $dq(v)$ would be observed in the time interval Δt, there being a total of $dq(v)\Delta t = NdG_m(v)$ vehicles. Hence

$$dq(v) = \frac{N\,dG_m(v)}{\Delta t} = \frac{Nv\,dG_m(v)}{L} = kv\,dG_m(v) = v\,dk(v) \qquad (II.24)$$

and over the whole

$$q = k\int_0^\infty v\,dG_m(v) = kE_m(v). \qquad (II.25)$$

In this equation q and k are related through the expected value of the instantaneous speed distribution (for more detail, see Sect. II.2.5.1). Now let us consider the local speed distribution, observed at a measuring point x. The probability that a vehicle is observed having a speed in the interval $(v,v+dv)$ is, since $dq(v) = q\,dG_1(v)$,

$$dG_1(v) = \frac{dq(v)}{q} = \frac{kv\,dG_m(v)}{kE_m(v)}$$

$$dG_1(v) = \frac{v}{E_m(v)}dG_m(v) \qquad (II.26)$$

[see Eq.(I.32)] and the expected value is

$$E_1(v) = \int_0^\infty v\,dG_1(v) = \frac{1}{E_m(v)}\int_0^\infty v^2dG_m(v).$$

Substituting the identity

$$\int_0^\infty v^2dG_m(v) = E_m(v^2) = \sigma_m^2 + [E_m(v)]^2$$

into the above equation (see Sect. II.1), one obtains

$$E_1(v) = E_m(v) + \frac{\sigma_m^2}{E_m(v)}. \qquad (II.27)$$

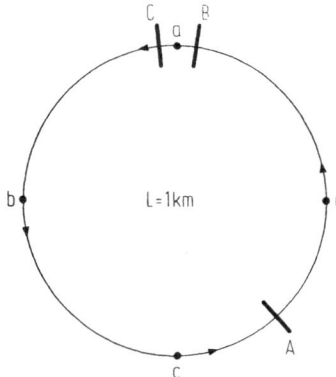

Fig. II.17

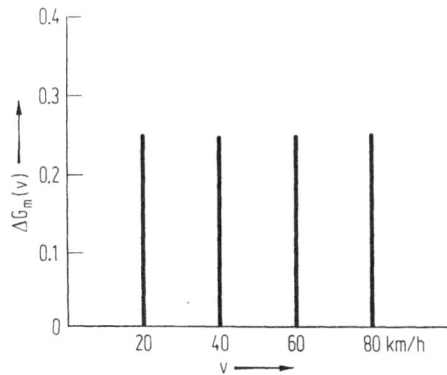

Fig. II.18

Except for the case in which all vehicles have the same speed ($\sigma_m = 0$), $E_1(v)$ is greater than $E_m(v)$. Equation (II.26) shows, how one can compute the local speed distribution from the instantaneous speed distribution. Under the assumption of homogeneous road conditions, the flow is stationary over time and distance.

Example 27. Four vehicles travel on the circular road (see Fig. II.17):

veh. a travels at 20 km/h,
veh. b travels at 40 km/h,
veh. c travels at 60 km/h,
veh. d travels at 80 km/h.

During an observation time of one hour the vehicles pass point A as follows:

20 times for $v_1 = 20$ km/h,
40 times for $v_2 = 40$ km/h,
60 times for $v_3 = 60$ km/h,
80 times for $v_4 = 80$ km/h.

Therefore, the total volume $q = (20 + 40 + 60 + 80) = 200$ veh/h, as observed locally; the resulting mean speed is

$$\bar{v}_l = \frac{1}{200} (20 \cdot 40 + 40 \cdot 40 + 60 \cdot 60 + 80 \cdot 80) = 60 \text{ km/h.}$$

Equation (II.26) allows $dG_1(v)$ [and also $E_1(v)$] to be computed without taking the observation time into account:

for veh. a: $\Delta G_m(20) = 0.25$,
for veh. b: $\Delta G_m(40) = 0.25$,
for veh. c: $\Delta G_m(60) = 0.25$,
for veh. d: $\Delta G_m(80) = 0.25$,

(see Fig. II.18).

It follows that

$$\Delta G_1(20) = \frac{v_i}{\bar{v}_m} \Delta G_m(v_i)$$

$$= \frac{20}{50} 0.25 = 0.1$$

$$\Delta G_1(40) = \frac{40}{50} 0.25 = 0.2$$

$$\Delta G_1(60) = \frac{60}{50} \cdot 0.25 = 0.3$$

$$\Delta G_1(80) = \frac{80}{50} \cdot 0.25 = 0.4$$

$$\overline{\sum_{i=1}^{4} \Delta G_1(v_i) = 1.0}$$

(Fig. II.19), and therefore

$$E_1(v) = \sum_{i=1}^{4} v_i \Delta G_1(v_i) = (0.1 \cdot 20 + 0.2 \cdot 40 + 0.3 \cdot 60 + 0.4 \cdot 80) = 60 \text{ km/h}.$$

Using Eq. (II.27), the expected value of the local speed can be calculated from the parameters of the instantaneous speed distribution

$$E_1(v) = E_m(v) + \frac{\sigma_m^2(v)}{E_m(v)}.$$

As shown in Sect. II.1,

$$\sigma_m^2(v) = E_m(v^2) - [E_m(v)]^2.$$

Computing

$$E_m(v^2) = \sum_{i=1}^{4} v_i^2 \Delta G_m(v_i) = 20^2 \cdot 0.25 + 40^2 \cdot 0.25 + 60^2 \cdot 0.25 + 80^2 \cdot 0.25$$

$$= 3000$$

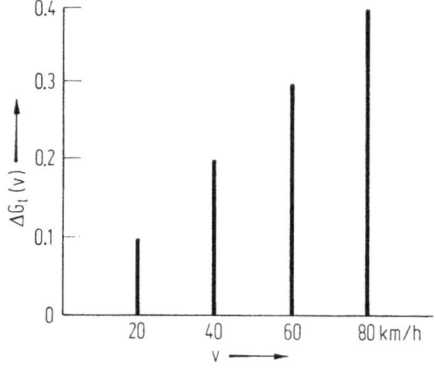

Fig. II.19

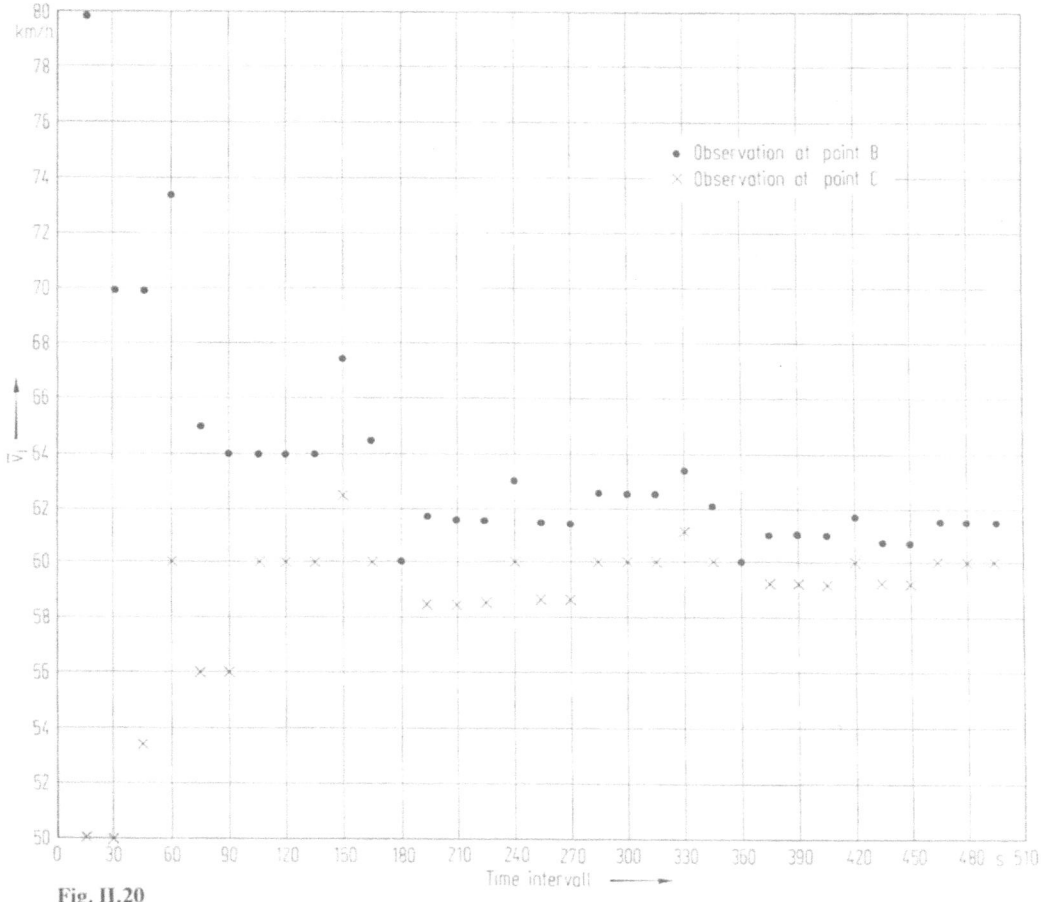

Fig. II.20

and with $E_m(v) = \bar{v}_m (= 50\,\text{km/h})$ we obtain

$$\sigma_m^2 = 3000 - 2500 = 500$$

so that

$$E_1(v) = 50 + \frac{500}{5} = 60\,\text{km/h}.$$

as above.

This example can also be used to illustrate the distinction between the expected value of a random variable and the mean of a sample which estimates that expected value (see Fig. II.20):

The larger the sample the more closely the observed sample mean $\bar{v}_1$ approaches the expected value $E_1(v)$. Further, the sequence of sample means, as the sample size increases, depends on the initial conditions of the observation process: in the upper half of the figure, the observation process begins with vehicle d (at point B), in the lower half with the vehicles a and d (at point C).

In addition it can be shown that the arithmetic mean of the instantaneous speed is equal to the harmonic mean of the local speed: from Eq. (II.24) we obtain $dq(v) = vdk(v)$ and from Eq.(II.26) $dq(v) = qdG_1(v)$. Setting these two expressions for $dq(v)$ equal yields

$$dk(v) = q\frac{1}{v}dG_1(v)$$

and hence

$$k = q\int_0^\infty \frac{1}{v}dG_1(v) = qE_1\left(\frac{1}{V}\right) \tag{II.28}$$

or with $1/v = w$ (see Sect. I.1.4)

$$k = qE_1(W). \tag{II.29}$$

But since, from Eq.(II.25),

$$\frac{k}{q} = \frac{1}{E_m(V)}$$

we obtain, for the population mean,

$$E_m(V) = \frac{1}{\int_0^\infty \frac{1}{v}dG_1(v)} = \frac{1}{E_1\left(\frac{1}{V}\right)} = \frac{1}{E_1(W)} \tag{II.30}$$

and for the sample mean,

$$\bar{v}_m = \frac{M}{\sum\limits_{i=1}^k \frac{1}{v_i}m_i} = \frac{M}{\sum\limits_{i=1}^k w_i m_i} = \frac{1}{\bar{w}_1}. \tag{II.31}$$

The reciprocity between $\bar{v}$ and $\bar{w}$ is therefore valid not just in the study of the kinematics of a single vehicle (cf. Sect. I.2.2).

For **Example 27** above we can therefore write

$$\bar{v}_m = \frac{200}{20\frac{1}{20} + 40\frac{1}{40} + 60\frac{1}{60} + 80\frac{1}{80}} = 50 \text{ km/h.}$$

One can evidently obtain Eq.(II.31) if one measures w_i directly instead of v_i. Let us define $F_m(w)$ as the probability distribution function for the instantaneously measured slowness, with expected value

$$E_m(W) = \int_\infty^0 w\, dF_m(w).$$

The sample mean is defined as

$$\bar{w}_m = \frac{1}{N}\sum_{i=1}^k w_i n_i = \sum_{i=1}^k w_i \Delta F_m(w_i).$$

Since, during the time interval $\Delta t_i = L/v_i = L w_i$ in our circular road example, all n_i vehicles having speed $v_i = 1/w_i$ will be observed just once, one observes at point x during the time T

$$m_i = \frac{T}{\Delta t_i} n_i = \frac{T}{L} \frac{n_i}{w_i} \qquad (\text{II}.32)$$

vehicles with slowness w_i.

Let us define $F_1(w)$ as the probability distribution function of the locally measured slowness, having expected value

$$E_1(W) = \int_{\infty}^{0} w \, dF_1(w).$$

The sample mean is defined as

$$\bar{w}_1 = \frac{1}{M} \sum_{i=1}^{k} w_i m_i.$$

In terms of our example, Eq. (II.32) can be used to obtain

$$M = \sum_{i=1}^{k} m_i = \frac{T}{L} \sum_{i=1}^{k} \frac{n_i}{w_i}$$

and the desired result

$$\bar{w}_1 = \frac{\sum\limits_{i=1}^{k} n_i}{\sum\limits_{i=1}^{k} \frac{n_i}{w_i}} = \frac{N}{\sum\limits_{i=1}^{k} n_i v_i} = \frac{1}{\bar{v}_m}.$$

Furthermore, corresponding to Eq. (II.26)

$$dF_m(w) = \frac{w}{E_1(W)} \cdot dF_1(w)$$

and corresponding to Eq. (II.26)

$$E_m(W) = E_1(W) + \frac{\sigma_1^2(W)}{E_1(W)}.$$

II.2.3.3 The Speed Distribution as Seen by a Moving Observer

The circular road example emphasized that the measured speed distribution depends on the method of measurement. When one takes local measurement at a fixed observation point on the circular road, high speeds are over-represented. The faster a vehicle travels, the more frequently it passes the measuring point during the observation period.

Let us now assume that the (absolute not relative) speeds are measured by an observer who himself travels at some constant speed v_0. It is clear that if $v_{min} < v_0 < v_{max}$, the resulting probability density function must go to zero at $v = v_0$ as sketched in Fig. II.21. Speeds of v_0 cannot be observed. Only speed $v < v_0$

(resulting from overtaking) or speeds $v > v_0$ (resulting from the observer being overtaken) be observed. The observed probability density function will be denoted as $g(v/v_0)$. Further, when the observer overtakes a vehicle, this is called an active overtaking; when the observer is overtaken, this is called a passive overtaking.

If the observer is stationary, i.e. $v_0 = 0$, all vehicles will overtake him. Let $dk(v) = k\,dG_m(v)$ be the density of that portion of the traffic stream in which all vehicles travel at speed v, so that, the observer measures on the average as in Eq. (II.24)

$$dq(v) = kv\,dG_m(v)$$

and, in total [Eq. (II.25)]

$$q = k \int_0^\infty v\,dG_m(v) = kE_m(v) \quad \text{[veh/time interval]}.$$

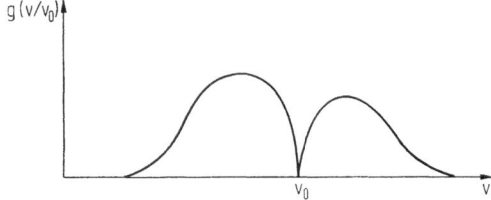

Fig. II.21

If the observer travels at speed v_0 he actively overtakes per time interval those vehicles whose speeds are $v < v_0$:

$$dq(v,v_0) = dq_p^a(v,v_0) = k(v_0 - v)\,dG_m(v)$$

$$q_p^a(v_0) = k \int_0^{v_0} (v_0 - v)\,dG_m(v).$$

(II.33)

He himself will be overtaken only by those vehicles whose speed is $v > v_0$:

$$dq(v,v_0) = dq_p^p(v,v_0) = k(v - v_0)\,dG_m(v)$$

$$q_p^p(v_0) = k \int_{v_0}^\infty (v - v_0)\,dG_m(v).$$

(II.34)

For $v_0 = 0$, Eq. (II.25) follows.

The total number of active and passive overtakings per time interval is then

$$q_p^{a+p}(v_0) = k\left[\int_0^{v_0} (v_0 - v)\,dG_m(v) + \int_{v_0}^\infty (v - v_0)\,dG_m(v) \right]$$

$$= k\left[v_0 \int_0^{v_0} dG_m(v) - \int_0^{v_0} v\,dG_m(v) + \int_{v_0}^\infty v\,dG_m(v) - v_0 \int_{v_0}^\infty dG_m(v) \right].$$

Since

$$\int_{v_0}^{\infty} dG_m(v) = 1 - \int_0^{v_0} dG_m(v)$$

$$\int_{v_0}^{\infty} v \, dG_m(v) = E_m(V) - \int_0^{v_0} v \, dG_m(v)$$

we obtain

$$q_p^{a+p}(v_0) = k \left\{ v_0 \int_0^{v_0} dG_m(v) - v_0 \left[1 - \int_0^{v_0} dG_m(v) \right] - \int_0^{v_0} v \, dG_m(v) + E_m(V) \right.$$

$$\left. - \int_0^{v_0} v \, dG_m(v) \right\}$$

$$= k \left[2v_0 \int_0^{v_0} dG_m(v) - v_0 + E_m(V) - 2 \int_0^{v_0} v \, dG_m(v) \right]$$

$$= k \left[E_m(V) - v_0 + 2 \int_0^{v_0} (v_0 - v) \, dG_m(v) \right]. \qquad (II.35)$$

With Eqs.(II.33) and (II.35) it is possible to calculate the probability that an observer, moving with speed v_0, observes a speed in the interval $(v, v + dv)$, where $v < v_0$. This probability is the quotient of the observed volume of vehicles which travel with a speed in the interval $(v, v + dv)$, divided by the total volume of all observed vehicles

$$dG(v|v_0 > v) = \frac{k(v_0 - v) \, dG_m(v)}{k \left[E_m(V) - v_0 + 2 \int_0^{v_0} (v_0 - v) \, dG_m(v) \right]}. \qquad (II.36)$$

Correspondingly, the probability that a speed in the interval $(v, v + dv)$, $v > v_0$ is observed, is:

$$dG(v|v_0 < v) = \frac{(v - v_0) \, dG_m(v)}{\left[E_m(V) - v_0 + 2 \int_0^{v_0} (v_0 - v) \, dG_m(v) \right]}. \qquad (II.37)$$

Finally, the complete density function of the observed speed is

$$g(v|v_0) = \frac{|v - v_0| g_m(v)}{\left[E_m(V) - v_0 + 2 \int_0^{v_0} (v_0 - v) \, dG_m(v) \right]}. \qquad (II.38)$$

Example 28. Let an observer moving with $v_0 = 30$ km/h be added to the vehicle stream in our "circular road" example of Sect. II.2.3.2. In order to calculate $\Delta G(v/30)$ we need the quantities $\bar{v}_m = E_m(V) = 50$ km/h, and

$$2 \cdot \sum_{v_i \leqq v_0} (v_0 - v_i) \, dG_m(v_i) = 2(30 - 20) 0.25 = 5$$

in which only the speed of vehicle a (20 km/h) enters, because it is smaller than v_0.

Thus

$$dG(20|30) = \frac{|30-20|\cdot 0.25}{50-30+5} = 0.1$$

$$dG(40|30) = \frac{|40-30|\cdot 0.25}{25} = 0.1$$

$$dG(60|30) = \frac{|60-30|\cdot 0.25}{25} = 0.3$$

$$dG(80|30) = \frac{|80-30|\cdot 0.25}{25} = 0.5$$

$$\overline{} \atop \Sigma = 1.0$$

(see Fig. II.22).

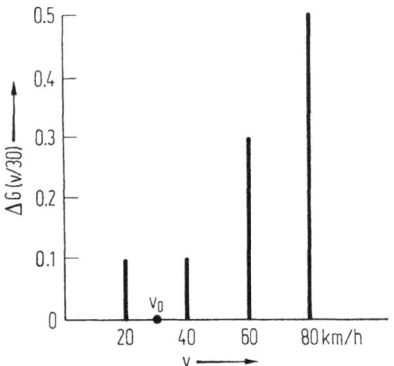

Fig. II.22

Instantaneous and local observations are obviously only two limiting cases of observation by a moving observer. Let us divide both the numerator and denominator of Eq. (II.36) by v_0:

$$dG(v|v_0 > v) = \frac{\left(1 - \dfrac{v}{v_0}\right) dG_m(v)}{\dfrac{E_m(V)}{v_0} - 1 + 2\displaystyle\int_0^{v_0}\left(1 - \dfrac{v}{v_0}\right) dG_m(v)}.$$

For an instantaneous observation, $v_0 = \infty$. This gives (after dividing by dv):

$$g(v|\infty) = \frac{g_m(v)}{-1 + 2\displaystyle\int_0^{\infty} dG_m(v)} = g_m(v).$$

For a local observation, $v_0 = 0$. Substituting $v_0 = 0$ into Eq. (II.37) yields the same result as in Eq. (II.26). Dividing by dv we obtain

$$g(v|0) = \frac{v g_m(v)}{E_m(V)} = g_l(v).$$

If the observer moves at a speed v_0 against the stream he encounters

$$B(v_0) = k \cdot \int_0^\infty (v_0 + v) \cdot dG_m(v)$$

$$= k \left[v_0 \cdot \int_0^\infty dG_m(v) + \int_0^\infty v \cdot dG_m(v) \right] \tag{II.39}$$

$$= k[v_0 + E_m(V)]$$

vehicles. For a stationary observer ($v_0 = 0$) we have

$$B(v_0) = k \cdot E_m(V) = q$$

[see Eq. (II.25)].

Example 29. In the ring example the observer drives at $v_0 = 30$ km/h against the stream. He encounters (given $k = 4$ and $E_m(V) = \bar{v}_m = 50$ km/h)

$$B(v_0) = 4(30 + 50) = 320 \text{ veh/h}.$$

II.2.3.4 Parameters of Speed Distributions for Non-constant Speeds

The assumption in the circular road example of unlimited overtaking possibilities restricts the transfer of these results to the very limited domain of freely flowing traffic. The assumption that every vehicle maintains a constant speed is generally unrealistic and only meaningful when this constant speed is interpreted as being the mean speed of that vehicle. The effects of the fluctuations in the speed of a single vehicle on the parameters of the speed distribution must be determined. Let (see Sect. I.2.1) $\hat{v}_t = u$ be the journey speed of one vehicle. The journey speeds of all vehicles are described by the distribution function $G_m(u)$ for the instantaneous speeds. Let the distribution of the speeds $(v)t$ of a single vehicle, measured over time, be described by $F_t(v|u)$. Further it is assumed on grounds of simplicity that all vehicles have this same speed distribution function[1]. The mean and variance of $v(t)$ are denoted by

$$\int_0^\infty v \, dF_t(v|u) = u \tag{II.40}$$

$$\int_0^\infty (v - u)^2 dF_t(v|u) = \bar{\sigma}_t^2. \tag{II.41}$$

The probability that at some point in time a vehicle is observed having a speed in the interval $(v, v + dv)$ is equal to the integral, over u, of the joint probability that that vehicle has, simultaneously, an overall journey speed in the interval $(u, u + du)$ and is travelling momentarily with a speed in the interval $(v, v + dv)$:

$$dG_m(v) = \int_{u=0}^\infty dF_t(v|u) dG_m(u). \tag{II.42}$$

It follows that

$$E_m(V) = \int_0^\infty v \, dG_m(v) = \int_{u=0}^\infty \left[\int_{v=0}^\infty v \, dF_t(v|u) \right] dG_m(u)$$

---

1 This assumption is also certainly not realistic, the actual pattern of speed variation is in general not known.

and with Eq. (II.40)

$$E_m(V) = \int_0^\infty u\, dG_m(u) = E_m(u).$$

(II.43)

Therefore, for the calculation of the mean value of the instantaneous speed, it makes no difference whether the speed fluctuations, defined in the same manner, are accounted for.

Example 30. Let $\Delta G_m(u)$ be (see Fig. II.23):

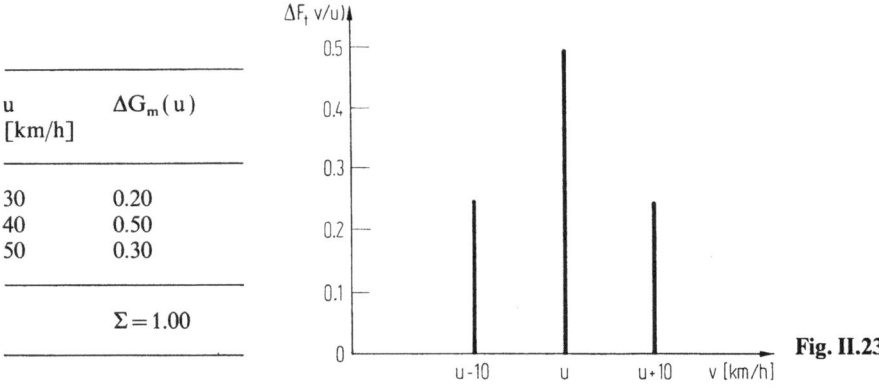

u [km/h]	$\Delta G_m(u)$
30	0.20
40	0.50
50	0.30
	$\Sigma = 1.00$

Fig. II.23

For every journey speed u the fluctuations have the same symmetrical distribution $\Delta F_t(v|u)$ (see Fig. II.24):

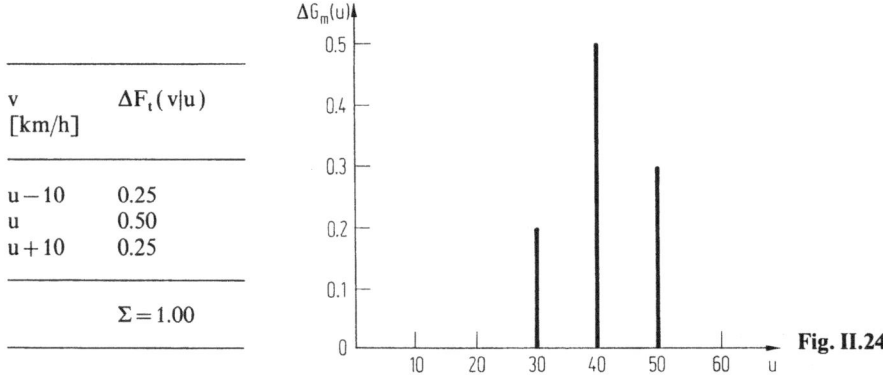

v [km/h]	$\Delta F_t(v|u)$
$u-10$	0.25
u	0.50
$u+10$	0.25
	$\Sigma = 1.00$

Fig. II.24

Rewriting Eq. (II.42) for a discrete distribution,

$$\Delta G_m(v) = \sum^u \Delta F_t(v|u)\, dG_m(u).$$

Therefore, the probability that $v = 20$ km/h, is (see Fig. II.25)

$$\Delta G_{m,v}(20) = \Delta G_{m,u}(30)\,\Delta F_t(20|30) + \Delta G_{m,u}(40)\,\Delta F_t(20|40)$$

$$+ \Delta G_{m,u}(50)\,\Delta F_t(20|50)$$

$$= 0.2\cdot0.25 + 0.5\cdot0.0 + 0.3\cdot0.0 = 0.05$$

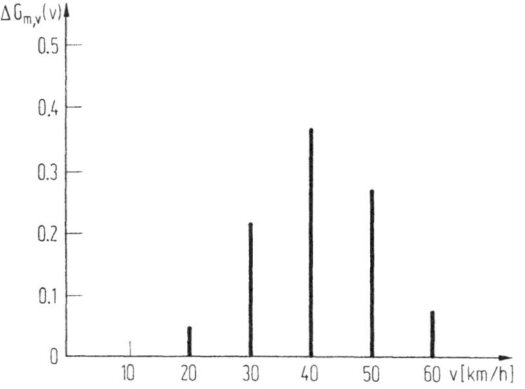

Fig. II.25

(see Fig. II.25). Correspondingly,

$$\Delta G_{m,v}(30) = 0.2 \cdot 0.5 \ +0.5 \cdot 0.25 + 0.3 \cdot 0.0 \ =0.225$$

$$\Delta G_{m,v}(40) = 0.2 \cdot 0.25 + 0.5 \cdot 0.5 \ +0.3 \cdot 0.25 = 0.375$$

$$\Delta G_{m,v}(50) = 0.2 \cdot 0.0 \ +0.5 \cdot 0.25 + 0.3 \cdot 0.5 \ =0.275$$

$$\Delta G_{m,v}(60) = 0.2 \cdot 0.0 \ +0.5 \cdot 0.0 \ +0.3 \cdot 0.25 = 0.075$$

$$\overset{v}{\sum} \Delta G_{m,v}(v) \ = 1.000$$

Using Figs. II.23 and II.24

$$E_m(V) \ = \overset{v}{\sum} v \, dG_{m,v}(v)$$

$$= 20 \cdot 0.5 + 30 \cdot 0.225 + 40 \cdot 0.375 + 50 \cdot 0.275 + 60 \cdot 0.075$$

$$= 41.0 \ \text{km/h}.$$

Equivalently,

$$E_m(U) = \overset{u}{\sum} u \, \Delta G_{m,u}(u) \ = 30 \cdot 0.2 + 40 \cdot 0.5 + 50 \cdot 0.3 = 41 \ \text{km/h}.$$

For other moments of the distribution, relationships similar to Eq. (II.43) do not hold. Define

$$\beta_n = \int\limits_0^\infty (v-u)^n dF_t(v|u)$$

as n-th central moment of the speed $v(t)$ of a single vehicle whose journey speed is u. For $n=0$

$$\beta_0 = 1 \cdot \int\limits_0^\infty dF_t(v|u)$$

for $n=1$

$$\beta_1 = \int_0^\infty (v-u)\,dF_t(v|u) = \int_0^\infty v\,dF_t(v|u) - u\int_0^\infty dF_t(v|u)$$

and, for $n=2$

$$\beta_2 = \int_0^\infty (v-u)^2\,dF_t(v|u) = \bar{\sigma}_t^2.$$

Similarly, define

$$\varepsilon_n = \int_0^\infty [v - E_m(V)]^n\,dG_{m,v}(v)$$

as the n-th central moment of the instantaneous speed distribution of all vehicles on a section of road. As before

$$\varepsilon_0 = 1; \quad \varepsilon_1 = 0; \quad \varepsilon_2 = \sigma_m^2.$$

Finally, the n-th central moment of the instantaneously measured journey speeds u of all vehicles is defined as

$$\delta_n = \int_0^\infty [u - E_m(U)]^n\,dG_{m,u}(u)$$

with

$$\delta_0 = 1; \quad \delta_1 = 0; \quad \delta_2 = \omega_m^2.$$

Just as $\bar{\sigma}_t^2$ was the variance of the speed v_t of a single vehicle about its journey speed u, so is the average value of the variance of the speeds of all vehicles taken about their respective journey speeds u defined as

$$\overline{\bar{\sigma}_t^2} = \int_0^\infty \bar{\sigma}_t^2\,dG_{m,u}(u) = \int_{u=0}^\infty \int_{v=0}^\infty (v-u)^2\,dF_t(v|u)\,dG_{m,v}(u)$$

$$= \int_0^\infty \left[\int_0^\infty v^2\,dF_t(v|u) - 2u_0\int_0^\infty v\,dF_t(v|u) + u^2\int_0^\infty dF_t(v|u)\right]dG_{m,v}(u).$$

With

$$\int_0^\infty v\,dF_t(v|u) = u \quad \text{and} \quad \int_0^\infty dF_t(v|u) = 1$$

we have

$$\overline{\bar{\sigma}_t^2} = \int_0^\infty \left[\int_0^\infty v^2\,dF_t(v|u) - u^2\right]dG_{m,u}(u)$$

$$= \int_0^\infty \int_0^\infty v^2\,dF_t(v|u)\,dG_{m,u}(u) - \int_0^\infty u^2\,dG_{m,u}(u).$$

Since Eq. (II.42) says

$$\int_0^\infty dF_t(v|u)\,dG_{m,u}(u) = dG_{m,v}(v),$$

it follows that

$$\int_0^\infty v^2 dF_t(v|u) dG_{m,u}(u) = v^2 dG_{m,v}(v)$$

and that

$$\int_0^\infty \int_0^\infty v^2 dF_t(v|u) dG_{m,u}(u) = \int_0^\infty v^2 dG_{m,v}(v) = E_m(V^2) = \sigma_m^2 + [E_m(V)]^2$$

(see Sect. II.1). In the same way,

$$\int_0^\infty u^2 dG_{m,u}(u) = E_m(U^2) = \omega_m^2 + [E_m(U)]^2.$$

Using Eq. (II.43) and the above results, we obtain

$$\overline{\sigma_t^2} = (\sigma_m^2 + [E_m(V)]^2) - (\omega_m^2 + [E_m(U)]^2) = \sigma_m^2 - \omega_m^2$$

or,

$$\sigma_m^2 = \omega_m^2 + \overline{\sigma_t^2}. \tag{II.44}$$

The average value of the variance of the speeds of all vehicles about their respective journey speeds $\overline{\sigma_t^2}$ is therefore smaller than the variance of the instantaneously measured speeds of all vehicles σ_m^2 by the quantity ω_m^2 which is the variance of the journey speeds of all vehicles. Further, the mean values of local speed distributions are not equal.
Recalling Eqs. (II.26) and (II.27),

$$dG_1(V) = \frac{v}{E_m(V)} dG_{m,v}(v)$$

$$E_1(V) = E_m(V) + \frac{\sigma_m^2}{E_m(V)}.$$

These can be substituted into Eqs. (II.43) and (II.44) to give

$$E_1(V) = E_m(U) + \frac{1}{E_m(U)}(\omega_m^2 + \overline{\sigma_t^2}) = E_m(U) + \frac{\omega_m^2}{E_m(U)} + \frac{\overline{\sigma_t^2}}{E_m(U)}$$

$$= E_1(U) + \frac{\overline{\sigma_t^2}}{E_m(U)}.$$

Finally, we quote without derivation the results

$$\omega_1^2 = \omega_m^2 \left[1 - \left[\frac{\omega_m}{E_m(U)}\right]^2\right] + \frac{\delta_3}{E_m(U)}$$

and

$$\sigma_1^2 = \sigma_m^2 \left[1 - \left[\frac{\sigma_m}{E_m(V)}\right]^2\right] + \frac{\varepsilon_3}{E_m(V)}.$$

In these equations δ_3 is the third central moment of the journey speeds, and ε_3 is the third central moment of the instantaneous speeds of all vehicles.

II.2.4 Headway Distributions

In Example 22 the Poisson distribution was used as an example of a discrete distribution. This distribution can be used to calculate the probability that m vehicles pass an observation point during a fixed time interval Δt, or equivalently, that at some instant n vehicles can be found instantaneously in some fixed distance Δx. Now let us introduce time or distance as a parameter thus obtaining a random process (see Fig. II.26). A random process is defined to include the set of all numbers that the random variable can assume over the range of the independent variables. In order to be able to describe traffic as a Poisson process (which is one type of random process), the following assumptions are required:[1]

1) The traffic stream must be stationary in the sense that $\lambda = $ constant; the probability that m vehicles appear in the interval $(t_0, t_0 + \Delta t)$ is independent of t_0:

$$P_{t_0, t + \Delta t}[M = m] = P_{\Delta t}[M = m].$$

2) The traffic stream has no memory; the record of past events provides no information as to the course of future events: $P_{t_0, t_0 + \Delta t}[M = m]$ is independent of the details of the process up to time t_0.

3) The simultaneous appearance of several vehicles at a location x_i can be neglected; i.e.

$$\lim_{\Delta t \to 0} \frac{P[M(x_i, t, \Delta t) > 1]}{\Delta t} = 0$$

(see Sect. II.2.1)

From these three conditions the probability $P_{\Delta t}[M = m]$ can be derived as

$$P_{\Delta t}[M = m] = \frac{(\lambda \Delta t)^m}{m!} e^{-\lambda \Delta t} \quad (m = 0, 1, 2, \ldots) \tag{II.45}$$

(or analogously, for distance

$$P_{\Delta x}[N = n] = \frac{(\varkappa \Delta x)^n}{n!} e^{-\varkappa \Delta x} \quad (n = 0, 1, 2, \ldots)),$$

and the distribution function as

$$P_{\Delta t}[M \leqq m] = \sum_{m_i \leqq m} \frac{(\lambda \Delta t)^{m_i}}{m_i!} e^{-\lambda \Delta t} \tag{II.46}$$

(see Example 22). The expected value is

$$E(M) = \lambda \Delta t$$

and the variance is

$$\sigma_M^2 = \lambda \Delta t.$$

Note that the mean and the variance are equal.

---

1 In the following, only events in time will be considered in general because the process defined over distance is completely analogous.

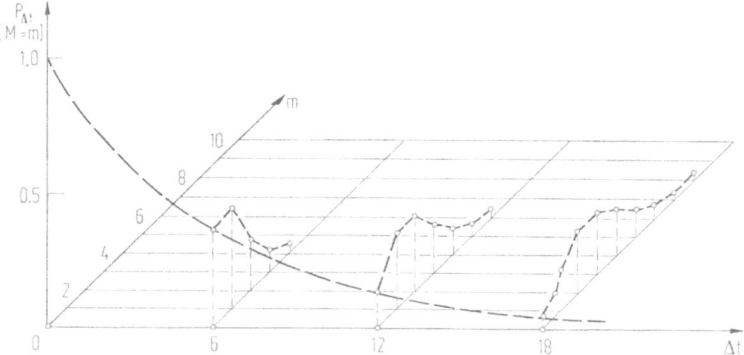

Fig. II.26. (From [35])

If the traffic stream is described by a Poisson process, then the time-headways (the time-headway is defined as the time interval between the passage of the same point, e.g. the rear bumper, of two successive vehicles measured at a fixed location) follow an exponential distribution. (The time-headway as defined here is often labelled simply as the headway, but the term headway is used by some writers to refer to the spacing between two vehicles. This latter quantity will, in this text, be labelled the distance-headway.)

This can be shown by noting that the probability that no vehicle appears ($m=0$) in an interval Δt is identical to the probability that a time-headway is $\geq \Delta t$

$$P_{\Delta t}[M=0] = \frac{(\lambda \Delta t)^0}{0!} e^{-\lambda \Delta t} = e^{-\lambda \Delta t}.$$

The above probability is just the complementary distribution function for an exponential random variable. The time-headway will be denoted by the continuous random variable Z:

$$P[Z>z] = e^{-\lambda z} \tag{II.47}$$

and the distance-headway by the random variable A,

$$P[A>a] = e^{-\varkappa a}.$$

Since $P[Z \leq \infty] = 1$ (see Sect. II.1) the probability that a vehicle has a time-headway $Z \leq z$ is given by

$$P[Z \leq z] = 1 - e^{-\lambda z}. \tag{II.48}$$

The corresponding probability density is

$$f(z) = \frac{dP[Z \leq z]}{dz} = \begin{cases} 0, & \text{for } z < 0 \\ \lambda e^{-\lambda z}, & \text{for } z \geq 0. \end{cases} \tag{II.49}$$

The expected value is $E(Z) = 1/\lambda$ and the variance is $\sigma_Z^2 = 1/\lambda^2$.

In general, exponential functions, when plotted on semi-logarithmic coordinates, are straight lines: for $z = 1/\lambda = E(Z)$ we have

$$P[Z > E(Z)] = e^{-\lambda/\lambda} = e^{-1} = 0.368.$$

As Fig. II.27 shows, the complementary distribution function can be plotted knowing only two points.

Thus it is especially easy to see from such a probability plot of the observed time-headways whether the process is stationary, at least to a first approximation (see Fig. II.28). Certainly a close fit of the observed time headways to a straight

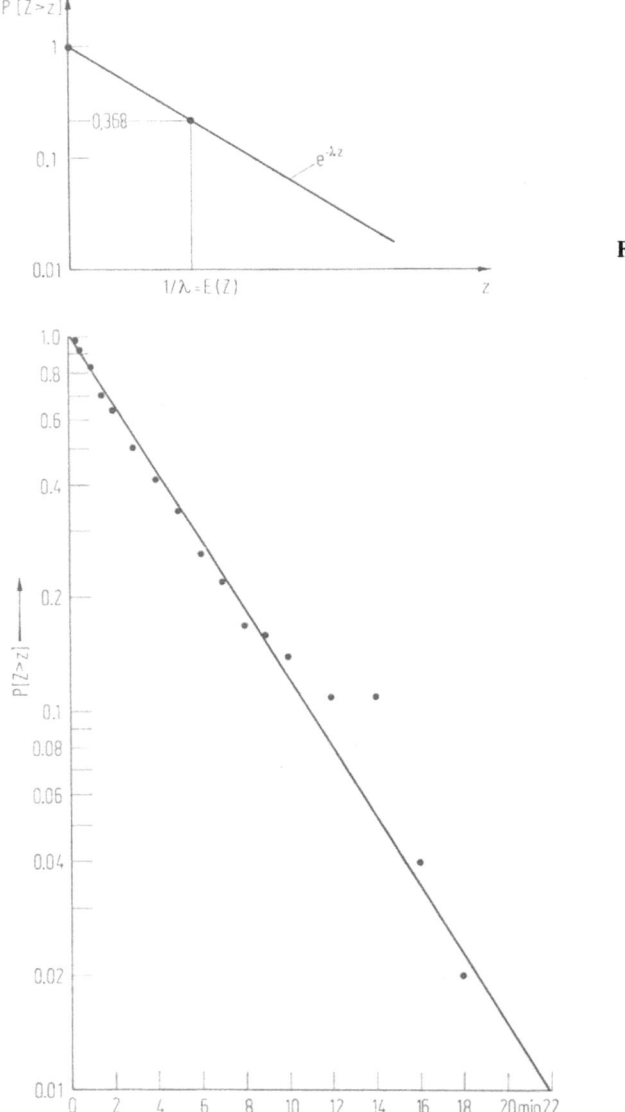

Fig. II.27

Fig. II.28. (From [34])

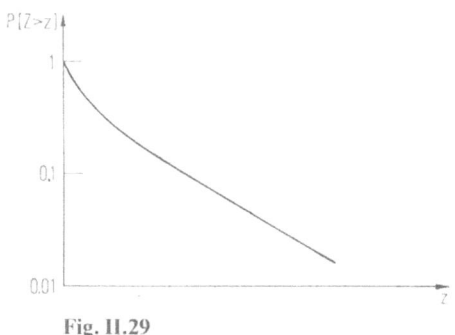

Fig. II.29

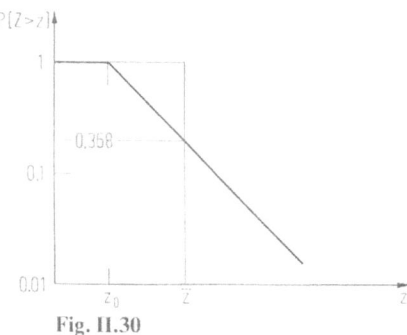

Fig. II.30

line is a necessary, but not sufficient condition for stationarity. A more reliable indicator for stationarity is the CUSUM function (see additionally Sect. II.2.1).

It is possible for free flowing traffic to be non-stationary implying $\lambda = \lambda(t)$ or $\varkappa = \varkappa(x)$ (see Sect. II.3.1).

Consider for example an observation period $T = T_1 + T_2$. During T_1 traffic is stationary with parameter λ_1 and, during T_2, with parameter λ_2. The probability of a time-headway $Z > z$, considering the entire observation period, is

$$P[Z>z] = \frac{T_1\lambda_1 e^{-\lambda_1 z} + T_2\lambda_2 e^{-\lambda_2 z}}{T_1\lambda_1 + T_2\lambda_2}. \tag{II.50}$$

This complementary distribution function is not a straight line when plotted on semi-logarithmic coordinates (Fig. II.29).

The generalization of Eq. (II.50) to a situation characterized by k successively different values of λ_i is

$$P[Z>z] = \frac{\sum\limits_{i=1}^{k} T_i\lambda_i e^{-\lambda_i z}}{\sum\limits_{i=1}^{k} T_i\lambda_i}. \tag{II.51}$$

In heavy traffic vehicles must frequently adjust their speeds to that of a vehicle in front, thus forming platoons (see Sect. II.3.3). If the vehicles had the same time-headway the complementary distribution function of the time-headway would be a step function (shown in Fig. II.30 by the light line). In reality, time-headways in platoons will be distributed somehow. If one assumes that these headways are likewise exponentially distributed, and that z_0 is the minimum time-headway then the complementary distribution function is (shown in Fig. II.30 by the dark line):

$$P[Z>z] = \begin{cases} 1, & \text{for } z < z_0 \\ e^{-\lambda'(z - z_0)}, & \text{for } z \geqq z_0. \end{cases} \tag{II.52}$$

The expected value from this distribution function must be equal to the actual mean time-headway $\bar{z} = 1/\lambda$. Thus we obtain

$$\lambda' = \frac{\lambda}{1 - z_0\lambda}.$$

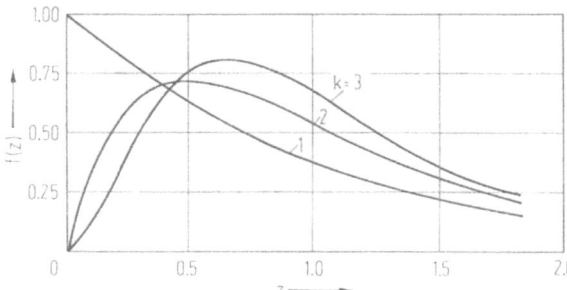

Fig. II.31. (From [36])

If it is assumed that the observed traffic results from the combination of two processes, one associated with those vehicles which travel freely, and the other associated with vehicles travelling in platoons, the resulting complementary distribution function is the superposition of the individual functions:

$$P[Z>z] = \frac{T_1\lambda_1}{T_1\lambda_1 + T_2\lambda_2} e^{-\lambda_1 z} + \frac{T_2\lambda_2}{T_1\lambda_1 + T_2\lambda_2} e^{-(\frac{\lambda_2}{1-z_0\lambda_2})(z-z_0)}. \qquad (II.53)$$

In order to account for the fact that very small headways in platoons occur seldom or not at all, the Erlang distribution

$$F(z) = P[Z \leq z] = \int_0^z \frac{(k\lambda)^k}{(k-1)!} y^{k-1} e^{-k\lambda y} dy \qquad (II.54)$$

is also frequently used. The corresponding probability density is characterized by having a maximum value — except for $k=1$ — at a value of $z>0$ (see Fig. II.31, where Cartesian coordinates are used).

As can be recognized from Fig. II.31 and also from Eq. (II.54), the Erlang distribution becomes the exponential distribution when $k=1$.

Since the Erlang distribution is defined for positive integer values of $k \geq 1$, it is a special case of the more general Pearson Type III distribution

$$P[Z \leq z] = \int_0^z \frac{h^k}{\Gamma(k)} y^{k-1} e^{-hy} dy \qquad (II.55)$$

with $h = \lambda k$, which holds for any positive value of k.

II.2.5 Lane Occupancy

Suppose that it is possible to measure the length of all the vehicles on a section of road. Then the magnitude

$$B = \left(\sum_{i=1}^N l_i \right) / X$$

is a measure of the traffic density, referred to as lane occupancy. Given that $\sum_{i=1}^{N} l_i = N \cdot l_f$, with l_f = mean vehicle length, we obtain

$$B = \frac{N \cdot l_f}{X} = l_f \cdot k. \tag{II.56}$$

Lane occupancy is, however, also

$$B = \left(\sum_{i=1}^{M} t_i \right) / T$$

where t_i is the time that a vehicle of length l_i and travelling at v_i occupies an observation point, $t_i = l_i / v_i$. Assuming that all vehicles possess the vehicle length, l_f, we obtain

$$\sum_{i=1}^{M} t_i = l_f \cdot \sum_{i=1}^{M} \frac{1}{v_i}.$$

From Eq. (II.31), however, we have

$$\sum_{i=1}^{M} \frac{1}{v_i} = \frac{M}{\bar{v}_m}$$

and hence

$$\Sigma t_i = \frac{l_f \cdot M}{\bar{v}_m}; \quad B = \frac{l_f \cdot M}{\bar{v}_m \cdot T}$$

or, taking $M/T = q$ and with $q / \bar{v}_m = k$

$$B = l_f \cdot k$$

as above.

Equation (II.56) can, however, also be derived from the earlier circular example. From Eq. (II.24), the subset of vehicles which pass the observation point at speed v is given by

$$dq(v) = k \cdot v \cdot dG_m(v).$$

When a vehicle of length l_f occupies the observation point for $\Delta t = l_f / v$, the occupancy resulting from all vehicles travelling at speed v is

$$dB(v) = dq(v) \cdot \Delta t = \frac{l_f \cdot k \cdot v \cdot dG_m(v)}{v} = l_f \cdot k \cdot dG_m(v)$$

and so the total occupancy is

$$B(v) = l_f \cdot k \cdot \int_0^{\infty} dG_m(v) = l_f \cdot k.$$

II.2.6 Relationship Between Parameters

II.2.6.1 Fundamental Considerations

In the circular road example the relationships between q, k and the expected speed were derived [Eqs. (II.25) and (II.29)]:

$$q = k\, E_m(V); \quad k = q\, E_1\left(\frac{1}{V}\right) = q E_1(W).$$

Similar relationships will be derived with the help of probability theory.

For a local measurement at a location x, the probability that a vehicle appears during the time interval $(t, t + dt)$ having speed v is equal to the product of the probability that a vehicle appears in $(t, t + dt)$ multiplied by the probability that that vehicle has speed v.

The probability that a vehicle appears at location x in dt follows from the definition of intensity and is $\lambda_x(t)\,dt$, while the probability that that vehicle has speed v is $g_1(v,x,t)\,dv = dG_1(v,x,t)$. The probability of both simultaneously occurring is

$$\lambda_x(t)\,dt\,dG_1(v,x,t).$$

Similarly the probability that a vehicle is located in the distance interval $(x, x + dx)$ at time t and having speed v is

$$\varkappa_t(x)\,dx\,dG_m(v,x,t).$$

The probability that a vehicle having speed v appears in the time interval dt at location x must be the same as the probability that that same vehicle is located in the distance interval $(x, x + v\,dt)$ at time t (see Fig. II.32).

Therefore,

$$\lambda_x(t)\,dG_1(v,x,t)\,dt = \varkappa_t(x)\,dG_m(v,x,t)\,v\,dt.$$

Integrating over all values of v, we obtain:

$$\lambda_x(t)\int_0^\infty dG_1(v,x,t) = \varkappa_t(x)\int_0^\infty v\,dG_m(v,x,t)$$

$$\lambda_x(t) = \varkappa_t(x)\,E_m[V(x,t)]$$

$$(II.57)$$

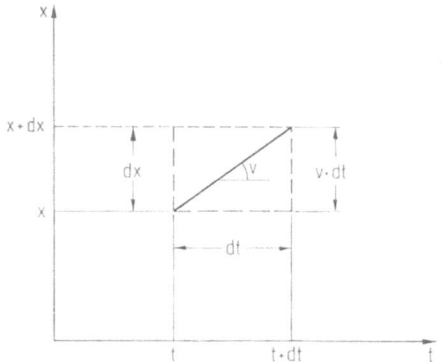

Fig. II.32

or transforming the above relationship from speed into slowness, we have

$$\lambda_x(t) \int_0^\infty \frac{1}{v} dG_1(v,x,t) = \varkappa_t(x) \int_0^\infty dG_m(v,x,t)$$

(II.58)

$$\lambda_x(t) E_1\left[\frac{1}{V(x,t)}\right] = \lambda_x(t) E_1[W(x,t)] = \varkappa_t(x).$$

Moreover the following forms are valid

$$q\left[\frac{vehicle}{time}\right] = k\left[\frac{vehicle}{distance}\right] \cdot v\left[\frac{distance}{time}\right]$$

or

$$k\left[\frac{vehicle}{distance}\right] = q\left[\frac{vehicle}{time}\right] \cdot w\left[\frac{time}{distance}\right]$$

because they are dimensionally consistent. The specific meaning of the particular parameters are determined by the measurement method.

II.2.6.2 The Effects of the Method of Measurement

II.2.6.2.1 Local Measurements and Measurement Intervals

A local measurement in a time interval $\Delta t = T$ measures some number M of vehicles and their speeds v_i independently of the state of the traffic stream. The traffic volume is $q = M/T$ as in Eq. (II.15), and the mean value of the local speed is

$$\bar{v}_1 = \frac{1}{M} \sum_{i=1}^M v_i = \frac{\sum_{i=1}^k v_i m_i}{M}.$$

The quantity

$$\bar{w}_1 = \frac{\sum_{i=1}^k \frac{1}{v_i} m_i}{M} = \frac{\sum_{i=1}^k w_i \cdot m_i}{M}$$

is labelled mean local slowness [see Eq. (II.31)]; the product $k = q\bar{w}_1$ is the traffic density.

If, as explained in Sect. II.2.1, local measurements are made at two locations x_0 and x_i, then, from the measurement at x_0,

$$q = \frac{\Phi_{x_0}(t_i) - \Phi_{x_0}(t_0)}{T} = \frac{M}{T}$$

and, from the two measurements together, using Eq. (II.17) with $\Delta x = X$, we have

$$k = \frac{\Phi_{x_i}(t_i) - \Phi_{x_0}(t_i)}{X} = \frac{N}{X}.$$

The quotient

$$\frac{q}{k} = \frac{M}{N}\frac{X}{T} = \bar{v}_m$$

is the slope of a secant in the fundamental diagram (see Fig. II.48); $\bar{v}_m$ is defined as the space mean speed. Let $z_i = t_i - t_{i+1}$ be the time-headway between two vehicles. If the measurement is carried out such that the beginning and the end of the measurement interval correspond to the appearance of a vehicle, then

$$T = \sum_i z_i^1$$

and it follows that

$$q = \frac{M}{T} = \frac{M}{\sum_{i=1}^{M} z_i} . \tag{II.59}$$

Because

$$k = q \cdot \bar{w}_1 = \frac{M}{T} \cdot \frac{\sum w_i}{M},$$

we have

$$k = \frac{\sum_1 w_i}{T} = \frac{\sum_i w_i}{\sum_i z_i}$$

$$\bar{w}_1 = k/q \tag{II.60}$$

as before. Let us assume that the measurement consists of several time intervals T_r. For each time interval, q_r, k_r and $\bar{v}_{mr}$ can be calculated from the above equations. Because $M = \sum_r M_r$ and $T = \sum_r T_r$, we obtain for the total measurement

$$q = \frac{M}{T} = \frac{\sum_r M_r}{\sum_r T_r} = \frac{\sum_r T_r q_r}{\sum_r T_r} \tag{II.61}$$

and setting

$$\sum_i \frac{1}{v_i} = \sum_i w_i = T_r k_r$$

we get

$$k = \frac{\sum_r \sum_i w_i}{T} = \frac{\sum_r T_r k_r}{\sum_r T_r} . \tag{II.62}$$

---

1 The upper limit of summation is M, because the sample of measured vehicles includes either the vehicle at the beginning or at the end of a time-headway.

From this we obtain

$$\bar{v}_m = \frac{q}{k} = \frac{\sum\limits_r T_r q_r}{\sum\limits_r T_r k_r}. \tag{II.63}$$

If all time intervals T_r are equally large, then

$$T = \sum_r T_r = r T_r$$

and thus

$$q = \frac{\sum\limits_r T_r q_r}{\sum\limits_r T_r} = \frac{T_r \sum\limits_r q_r}{r T_r} = \frac{\sum\limits_r q_r}{r}. \tag{II.64}$$

Correspondingly,

$$k = \frac{\sum\limits_r T_r k_r}{\sum\limits_r T_r} = \frac{\sum\limits_r k_r}{r} \tag{II.65}$$

and

$$\bar{v}_m = \frac{q}{k} = \frac{\sum\limits_r q_r}{\sum\limits_r k_r}. \tag{II.66}$$

Partitioning the measurement such that the number of vehicles M_r are observed in each interval, we have

$$q = \frac{\sum\limits_r M_r}{\sum\limits_r T_r} = \frac{r M_r}{\sum\limits_r T_r}$$

or with

$$T_r = \frac{M_r}{q_r}$$

$$q = \frac{r M_r}{M_r \sum\limits_r \dfrac{1}{q_r}} = \frac{r}{\sum\limits_r \dfrac{1}{q_r}} \tag{II.67}$$

$$k = \frac{\sum\limits_r T_r k_r}{\sum\limits_r T_r} = \frac{\sum\limits_r \dfrac{M_r}{q_r} k_r}{\sum\limits_r \dfrac{M_r}{q_r}} = \frac{M_r \sum\limits_r \dfrac{k_r}{q_r}}{M_r \sum\limits_r \dfrac{1}{q_r}}$$

$$= \frac{\sum\limits_r \dfrac{k_r}{q_r}}{\sum\limits_r \dfrac{1}{q_r}} = \frac{\sum\limits_r \bar{w}_{1_r}}{\sum\limits_r \dfrac{1}{q_r}} \tag{II.68}$$

$$\bar{v}_m = \frac{q}{k} = \frac{r}{\sum\limits_r \dfrac{k_r}{q_r}} = \frac{r}{\sum\limits_r \bar{w}_{1_r}} = \frac{1}{\bar{w}_1}. \tag{II.69}$$

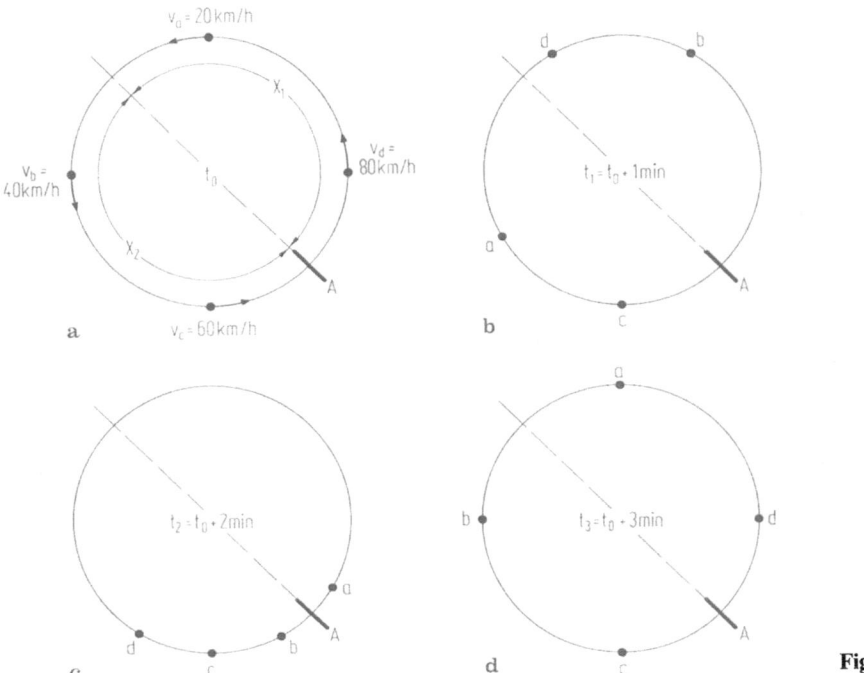

Fig. II.33a−d

Example 31. Let the vehicles in Example 27 in Sect. II.2.3.2 be located, at time t_0, at the position shown in Fig. II.33a. Each minute they travel a distance $\Delta x_i = v_i \cdot 1\,000/60$ min.
Therefore:

veh. a at 20 km/h; $\Delta x_1 = 16.6\overline{6}\ v_i = \ \ 333.\overline{3}$ m
veh. b at 40 km/h; Δx_2 $\qquad = \ \ 666.\overline{6}$ m
veh. c at 60 km/h; Δx_3 $\qquad = 1000.0$ m
veh. d at 80 km/h; Δx_4 $\qquad = 1333.\overline{3}$ m.

The positions of the vehicles are shown at one minute interval in Figs. II.33b − d. Consider two observation intervals: $T_1 = t_1 - t_0 = 1$ min and $T_2 = t_3 - t_1 = 2$ min. During T_1 vehicles b, c and d pass location A just once: $M_1 = 3$ veh/min and $q_1 = 3$ veh/min. Passing location A during T_2 are

veh. b − 1 time
veh. b − 1 time
veh. c − 2 times
veh. d − 3 times,

and in total $M_2 = 7$ veh, so that $q_2 = 7$ veh/2 min. In a total of 3 min, $M = 10$ veh will be observed:

$q = 10$ veh/3 min

or from Eq. (II.61)

$$q = \frac{\frac{3}{1} \cdot 1 + \frac{7}{2} \cdot 2}{1 + 2} = \frac{10}{3} \text{ veh/min.}$$

Since in the example, k_r equals 4 veh/km and is thus a constant independent of the time interval, the calculation of k is trivial. Applying Eq. (II.63), the space mean speed is computed as

$$\bar{v}_m = \frac{q}{k} = \frac{10/3}{4} = \frac{5}{6} \text{ km/min} = 50 \text{ km/h.}$$

Let observation be made during r = 2 equal time intervals:

$$T_1 = t_2 - t_1 = 1 \text{ min} \quad \text{and} \quad T_2 = t_3 - t_2 = 1 \text{ min.}$$

During T_1 the vehicles a, c and d pass location A just once; $M_1 = 3$ veh and $q_1 = 3$ veh/1 min. During T_2 4 vehicles (b, c and twice d) pass the location A, so that $M_2 = 4$ veh and $q_2 = 4$ veh/1 min. In the total time of 2 min, M = 7 veh are observed. Thus, q = 7 veh/2 min, or from Eq. (II.64)

$$q = \frac{\frac{3}{1} \cdot 1 + \frac{4}{1} \cdot 1}{1 + 1} = \frac{7}{2} \text{ veh/min.}$$

Here again $k_r = 4$ veh/km is a constant. From Eq. (II.66), v_m is calculated as

$$\bar{v}_m = \frac{q}{k} = \frac{7}{8} \text{ km/min} = 52 \text{ km/h.}$$

Deviations from $\bar{v}_m = 50$ km/h, such as appear in the preceding calculation, are discussed in connection with Fig. II.20.

II.2.6.2.2 Instantaneous Measurement and Measurement Intervals

A local measurement over a distance interval $\Delta x = X$ measures some number N of vehicles and (theoretically) their speeds v_i independently of the state of the traffic stream. From Eq. (II.20) k = N/X is the traffic density and the mean value of the instantaneously measured speeds is

$$\bar{v}_m = \frac{1}{N} \sum_{i=1}^{N} v_i = \frac{\sum_{i=1}^{k} n_i v_i}{N}.$$

The product $k\bar{v}_m = q$ is labelled the traffic volume. If, as explained in Sect. II.2.2, instantaneous measurements are made at two time points t_0 and t_1, then, from the measurement at t_0

$$k = \frac{\Psi_{t_0}(x_i) - \Psi_{t_0}(x_0)}{X} = \frac{N}{X}$$

and, from the two measurements together, using Eq. (II.21) with $\Delta t = T$, we have

$$q = -\frac{\Psi_{t_1}(x_i) - \Psi_{t_0}(x_i)}{T} = \frac{M}{T}.$$

The quotient

$$\frac{q}{k} = \frac{M}{N}\frac{X}{T} = \bar{v}_m$$

will, as in Sect. II.2.5.2.1, be labelled the space mean speed. Let $a_i = x_i - x_{i+1}$ be the distance-headway between two vehicles. If the measurement is carried out such that the beginning and the end of the measurement interval X correspond to the location of a vehicle, then

$$X = \sum_i a_i \quad [1]$$

and it follows that

$$k = N/X = N / \sum_{i=1}^{N} a_i . \tag{II.70}$$

Because $q = k \cdot \bar{v}_m$, we have

$$q = \frac{\sum_i v_i}{N} \cdot \frac{N}{X} = \frac{\sum_i v_i}{X} = \frac{\sum_i v_i}{\sum_i a_i} . \tag{II.71}$$

Hence

$$\bar{v}_m = q/k$$

as before. Let us assume that the measurement includes several space intervals X_r. For each space interval, k_r, q_r, and $\bar{v}_{m_r}$ can be calculated from the above equations. With

$$N = \sum_r N_r \quad \text{and} \quad X = \sum_r X_r$$

we obtain for the total measurement

$$k = \frac{N}{X} = \frac{\sum_r N_r}{\sum_r X_r} = \frac{\sum_r X_r k_r}{\sum_r X_r} \tag{II.72}$$

and setting

$$v_i = X_r q_r$$

yields

$$q = \frac{\sum_r \sum_i v_i}{X} = \frac{\sum_r X_r q_r}{\sum_r X_r} . \tag{II.73}$$

---

1 The upper limit of the summation is N, because the sample of measured vehicles includes either the vehicle at the beginning or at the end of a distance-headway.

From this we obtain

$$\bar{v}_m = \frac{q}{k} = \frac{\sum_r X_r q_r}{\sum_r X_r k_r} . \tag{II.74}$$

If all space intervals X_r are equally large, then

$$x = \sum_r X_r = r X_r$$

and thus

$$k = \frac{\sum_r X_r k_r}{\sum_r X_r} = \frac{X_r \sum_r k_r}{r X_r} = \frac{\sum_r k_r}{r} . \tag{II.75}$$

Correspondingly,

$$q = \frac{\sum_r X_r q_r}{\sum_r X_r} = \frac{\sum_r q_r}{r} \tag{II.76}$$

and

$$\bar{v}_m = \frac{q}{k} = \frac{\sum_r q_r}{\sum_r k_r} . \tag{II.77}$$

Partitioning the measurement so that the same number of vehicles N_r are observed in each space interval, we have

$$k = \frac{\sum_r N_r}{\sum_r X_r} = \frac{r N_r}{\sum_r X_r}$$

or, with $X_r = N_r / k_r$,

$$k = \frac{r N_r}{N_r \sum_r \frac{1}{k_r}} = \frac{r}{\sum_r \frac{1}{k_r}} \tag{II.78}$$

$$q = \frac{\sum_r X_r q_r}{\sum_r X_r} = \frac{\sum_r \frac{N_r}{k_r} q_r}{\sum_r \frac{N_r}{k_r}} = \frac{N_r \sum_r \frac{q_r}{k_r}}{N_r \sum_r \frac{1}{k_r}} = \frac{\sum_r \bar{v}_{m_r}}{\sum_r \frac{1}{k_r}} \tag{II.79}$$

$$\bar{v}_m = \frac{q}{k} = \frac{\sum_r \frac{q_r}{k_r}}{r} = \frac{\sum_r \bar{v}_{m_r}}{r} . \tag{II.80}$$

Example 32. Consider again the traffic flow process for the circular road as shown in Figs. II.33a – d. At time t_1 measurements are made on the two road sections X_1 and X_2, each being 500 m in length. Then, $k_1 = k_2 = 2$ veh/0.5 km and, from Eq. (II.71)

$$q_1 = \frac{v_b + v_d}{X_1} = \frac{40 + 80}{0.5} = 240 \text{ veh/h}$$

$$q_2 = \frac{v_a + v_c}{X_2} = \frac{20 + 60}{0.5} = 160 \text{ veh/h}$$

from which

$$k = \frac{k_1 + k_2}{2} = 2 \text{ veh/0.5 km}$$

$$q = \frac{q_1 + q_2}{2} = \frac{400}{2} = 200 \text{ veh/h}$$

$$\bar{v}_m = \frac{q}{k} = \frac{200}{2 \cdot 2} = 50 \text{ km/h}.$$

Table 1 collects all formulas developed so far.

II.2.6.2.3 Quasi-local Measurements

In contrast to the local measurement method as rigorously defined, the usual technique is to measure the travel time over a comparatively short distance Δx between two detectors. Such measurements are called quasi-local.

Table 1

	Local						Instantaneous						
				Several measurements						Several measurements			
	T variable	T fixed	Single measurement	General	Fixed time intervals	Fixed number of vehicles	X variable	X fixed	Single measurement	General	Fixed distance intervals	Fixed number of vehicles	
q	$\dfrac{M}{T}$	$\dfrac{M}{\sum^i z_i}$	$\dfrac{M}{T}$	$\dfrac{\sum^r T_r q_r}{\sum^r T_r}$	$\dfrac{\sum^r q_r}{r}$	$\dfrac{r}{\sum^r (1/q_r)}$	$\dfrac{\sum^i v_i}{X}$	$\dfrac{\sum^i v_i}{\sum^i a_i}$	$\dfrac{\sum^i \Delta x_i}{X \Delta t}$	$\dfrac{\sum^r X_r q_r}{\sum^r X_r}$	$\dfrac{\sum^r q_r}{r}$	$\dfrac{\sum^r (q_r/k_r)}{\sum^r 1/k_r}$
k	$\dfrac{\sum^i w_i}{T}$	$\dfrac{\sum^i w_i}{\sum^i z_i}$	$\dfrac{\sum^i \Delta t_i}{T \Delta x}$	$\dfrac{\sum^r T_r k_r}{\sum^r T_r}$	$\dfrac{\sum^r k_r}{r}$	$\dfrac{\sum^r (k_r/q_r)}{\sum^r (1/q_r)}$	$\dfrac{N}{X}$	$\dfrac{N}{\sum^i a_i}$	$\dfrac{N}{X}$	$\dfrac{\sum^r X_r k_r}{\sum^r X_r}$	$\dfrac{\sum^r k_r}{r}$	$\dfrac{r}{\sum^r (1/k_r)}$
$\bar{v}_m$	$\dfrac{M}{\sum^i w_i}$		$\dfrac{M \Delta x}{\sum^i \Delta t_i}$	$\dfrac{\sum^r T_r q_r}{\sum^r T_r k_r}$	$\dfrac{\sum^r q_r}{\sum^r k_r}$	$\dfrac{r}{\sum^r (k_r/q_r)}$	$\dfrac{\sum^i v_i}{N}$		$\dfrac{\sum^i \Delta x_i}{N \Delta t}$	$\dfrac{\sum^r X_r q_r}{\sum^r X_r k_r}$	$\dfrac{\sum^r q_r}{\sum^r k_r}$	$\dfrac{\sum^r (q_r/k_r)}{r}$

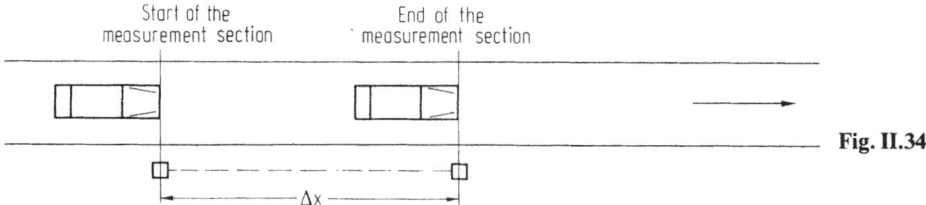

Fig. II.34

At both measuring points one obtains $q = M/T$. The quantity $v_i = \Delta x/\Delta t_i$ is, by definition, a journey speed. Nevertheless we label

$$\bar{v}_l = \frac{1}{M} \sum_i v_i = \frac{1}{M} \sum_i \frac{\Delta x}{\Delta t_i} \tag{II.81}$$

as the average local speed or time-mean-speed. The quantity

$$\bar{v}_m = \frac{M}{\sum_i \frac{1}{v_i}} = \frac{M}{\sum_i \frac{\Delta t_i}{\Delta x}} = \frac{M\Delta x}{\sum_i \Delta t_i} = \frac{\Delta x}{\overline{\Delta t}} \tag{II.82}$$

denotes the average instantaneous speed or space-mean-speed, and

$$\bar{w}_l = \frac{1}{\bar{v}_m} = \frac{\sum_i \Delta t_i}{M\Delta x} = \frac{\overline{\Delta t}}{\Delta x} \tag{II.83}$$

is the average local slowness. These results allow the calculation of the traffic density as

$$k = q\bar{w}_l = \frac{\sum_i \frac{1}{v_i}}{T} = \frac{\sum_i w_i}{T} = \frac{\sum_i \Delta t_i}{T\Delta x} = \frac{M\overline{\Delta t}}{T\Delta x}. \tag{II.84}$$

II.2.6.2.4 Quasi-instantaneous Measurements

Instantaneous speeds of vehicles on a road are in practice impossible to measure. Therefore, in general, two aerial photos separated by a comparatively short time-interval Δt must suffice. Such measurements are called quasi-instantaneous.

From such a quasi-instantaneous measurement (Fig. II.35) is obtained $k = N/X$. Here also the journey speed over the observation interval

$$v_i = \Delta x_i/\Delta t$$

is set equal to the instantaneous speed and called the average instantaneous speed or space-mean-speed:

$$\bar{v}_m = \frac{1}{N} \sum_i v_i = \frac{\sum_i \Delta x_i}{N\Delta t} = \frac{\overline{\Delta x}}{\Delta t}. \tag{II.85}$$

The product

$$q = k\bar{v}_m = \frac{\sum_i v_i}{X} = \frac{\sum_i \Delta x_i}{X\Delta t} = \frac{N\overline{\Delta x}}{X\Delta t} \tag{II.86}$$

is again the traffic volume.

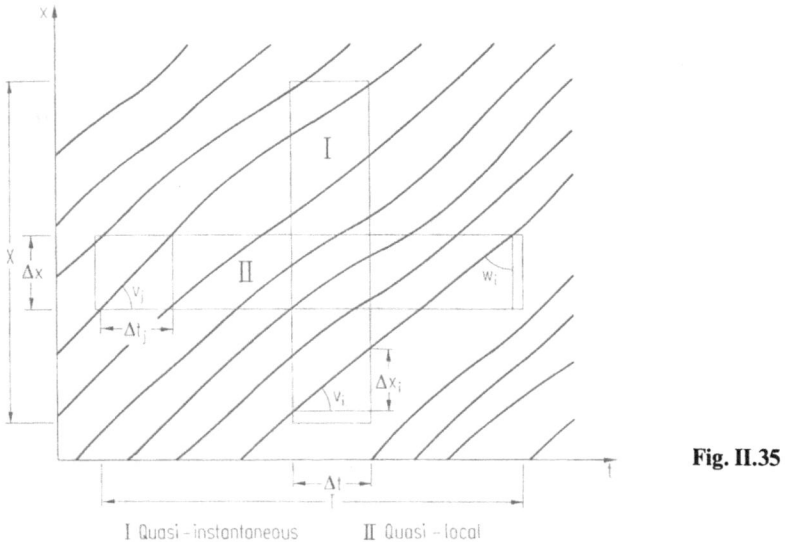

Fig. II.35

I Quasi-instantaneous   II Quasi-local

II.2.6.2.5 The Generalized Relationship

Repeating Eq. (II.86), $q = \left(\sum_i \Delta x_i \right) / X \Delta t$. The denominator corresponds to the portion of the x-t plane covered by the quasi-instantaneous measurement (see Fig. II.35); the numerator corresponds to the sum of the distances covered by the N observed cars in the indicated portion of the x-t plane.

For a local measurement, $q = M/T$. Multiplying numerator and denominator by Δx, we obtain quantities analogous to those above:

$$q = M \, \Delta x / T \, \Delta x.$$

With a quasi-local measurement, we know from Eq. (II.84) that

$$k = \left(\sum_i \Delta t_i \right) / T \, \Delta x.$$

Here also the denominator corresponds to the portion of the x-t plane covered by the quasi-local measurement (see Fig. II.35), and the numerator to the total time spent by the M observed vehicles in this same portion of the x-t plane.

For an instantaneous observation, $k = N/X$. Multiplying numerator and denominator by Δt, $k = N \cdot \Delta t / X \cdot \Delta t$ we obtain quantities which have the same meaning for the quasi-local measurement. If A is the area of some arbitrary portion of the x-t plane then, since

$$M \, \Delta x = \sum_i x_i \quad \text{and} \quad N \, \Delta t = \sum_i t_i$$

we can define in general:

$$q = \left(\sum_i \Delta x_i \right) / A \tag{II.87}$$

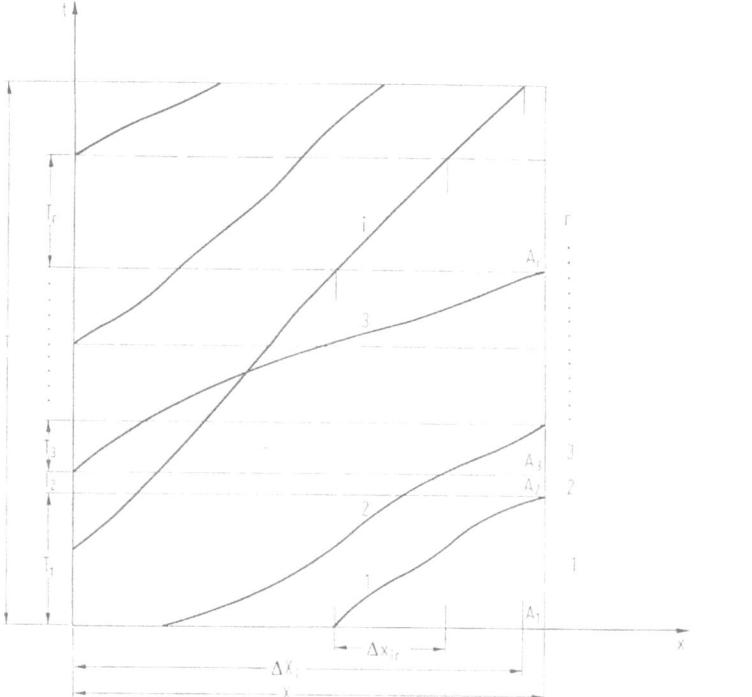

Fig. II.36

$$k = \left(\sum_i \Delta t_i \right) \Big/ A \qquad\qquad\qquad (II.88)$$

$$\bar{v}_m = q/k = \left(\sum_i \Delta x_i \right) \Big/ \left(\sum_i \Delta t_i \right). \qquad\qquad (II.89)$$

In the formulas for the quasi-local and quasi-instantaneous measurements, Δt and Δx can be chosen arbitrarily large, subject to the condition that the following equations remain valid:

$$M\Delta x = \sum_i \Delta x_i \quad \text{and} \quad N \Delta t = \sum_i \Delta t_i.$$

That is, all vehicles in A either traverse the entire distance Δx or remain in A for the entire time Δt. If several measurements are carried out, subject to the above conditions, and if, for the different portions A_r, of the x-t plane, we set

$$\Delta x = X_r \quad \text{or} \quad \Delta t = T_r,$$

then we have

$$A_r = TX_r = XT_r$$

and (see Fig. II.36)

$$\sum_r X_r = X; \quad \sum_r T_r = T; \quad \sum_r A_r = XT.$$

The distance travelled by vehicle i in A_r will be denominated by Δx_{ir}, and the time spent by vehicle i in A_r denominated by Δt_{ir}. Then Eq. (II.87) can be rewritten as

$$\sum_i \Delta x_{ir} = q_r A_r.$$

The distance travelled by vehicle i in $A = \sum^r A_r$ is denoted by ΔX_i, and the time spent, by ΔT_i. Thus

$$\sum_r \Delta x_{ir} = \Delta X_i \quad \text{or} \quad \sum_r \Delta t_{ir} = \Delta T_i.$$

From this we can write

$$\sum_r \sum_i x_{ir} = \sum_r q_r A_r = \sum_i \Delta X_i \tag{II.90}$$

and, proceeding similarly from Eq. (II.88), we obtain

$$\sum_r \sum_i \Delta t_{ir} = \sum_r k_r A_r = \sum_i \Delta T_i. \tag{II.91}$$

Rewriting Eq. (II.61)

$$q = \left(\sum_r T_r q_r \right) \Big/ \left(\sum_r T_r \right)$$

then multiplying numerator and denominator by X,

$$q = \left(\sum_r A_r q_r \right) / A$$

and with Eq. (II.90)

$$q = \left(\sum_i \Delta X_i \right) / A. \tag{II.92}$$

Correspondingly after multiplying numerator and denominator by T Eq. (II.72) becomes

$$k = \left(\sum_r A_r k_r \right) / A$$

and using Eq. (II.91)

$$k = \left(\sum_i \Delta T_i \right) / A. \tag{II.93}$$

The definitions for q and k which were derived in Sects. II.2.6.2.3 and II.2.6.2.4 for quasi-local and quasi-instantaneous measurements are therefore valid for arbitrarily large sections $A = XT$ of the x-t plane without accounting for the conditions defined for the subsections A_r of A. These condition require that the trajectory of a vehicle traverses either the entire length or the entire width of the designated area. The definitions of q and k are also independent of whether the traffic flow within the designated section A is stationary or non-stationary. As with the subsections A_r, the average instantaneous speed of the traffic stream in A is defined as

$$\bar{v}_m = \left(\sum_i \Delta X_i \right) \Big/ \left(\sum_i \Delta T_i \right). \tag{II.94}$$

Example 33. Figure II.37 shows a section of a measurement on a two-lane rural road. The trajectories are approximated by straight lines, whose slopes correspond to the journey speeds. The traffic volume q, the traffic density k, and the average speed $\bar{v}_m$ are to be computed.

From Eq. (II.92) taking $A = X \cdot T = 500 \text{ m} \cdot 100 \text{ s} = 50\,000 \text{ ms}$ we derive

$$q = \left(\sum_i \Delta X_i \right) / A = 3\,020 \text{ vehm}/50\,000 \text{ ms} = 0.0604 \text{ veh/s} = 217.44 \text{ veh/h}.$$

From Eq. (II.93)

$$k = \left(\sum_i \Delta X_i \right) / A = 230 \text{ vehs}/50\,000 \text{ ms} = 0.0046 \text{ veh/m} = 4.6 \text{ veh/km}$$

and from Eq. (II.94)

$$\bar{v}_m = \left(\sum_i X_i \right) \Big/ \left(\sum_i T_i \right) = 3\,020/230 = 13.13 \text{ m/s} = 47.27 \text{ km/h}.$$

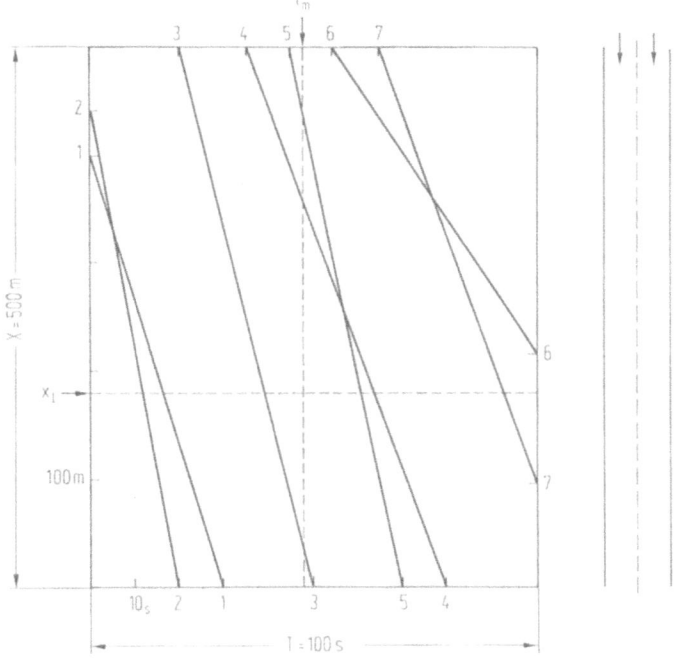

Fig. II.37

These values for q and k should be compared with those from either a local or an instantaneous measurement.

From a local measurement at point x_1 over time interval $T = 100$ s the local traffic flow

$$q_{x_1} = 6 \text{ veh}/100 \text{ s} = 0.06 \text{ veh/s} = 216 \text{ veh/h}$$

is obtained. From an instantaneous measurement at time t_m the traffic density

$$k_{t_m} = 3 \text{ veh}/500 \text{ m} = 0.006 \text{ veh/m} = 6 \text{ veh/km}$$

is derived.

Consider a platoon of $M + 1$ vehicles traversing a distance x (Fig. II.38); replace the trajectories of the first and last vehicle with straight lines whose slopes are the journey speeds $\hat{v}_0$ and $\hat{v}_M$ resprectively. Then

$$A = TX - \frac{X^2}{2}\left(\frac{1}{\bar{v}_0} + \frac{1}{\bar{v}_m}\right) = TX - \frac{X^2}{2}(\bar{w}_0 + \bar{w}_m)$$

and furthermore noting the footnote on page 76

$$q = \frac{\Sigma \Delta X_i}{A} = \frac{MX}{TX - \dfrac{X^2}{2}(\bar{w}_0 + \bar{w}_M)} = \frac{M}{T - \dfrac{X}{2}(\bar{w}_0 + \bar{w}_M)}$$

[if $X = 0$, then $q = M/T$ as in Eq. (II.59)]

$$k = \frac{\sum_i \Delta T_i}{A} = \frac{\sum_i X\bar{w}_i}{TX - \dfrac{X^2}{2}(\bar{w}_0 + \bar{w}_M)} = \frac{\sum_i \bar{w}_i}{T - \dfrac{X}{2}(\bar{w}_0 + \bar{w}_M)}$$

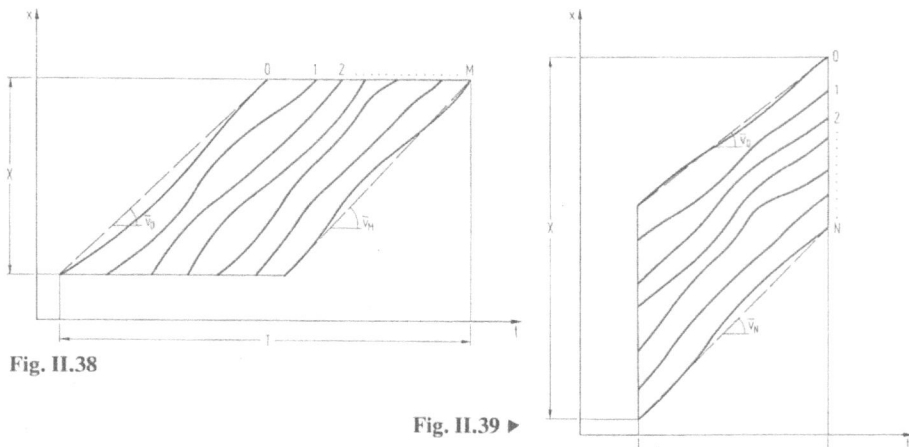

Fig. II.38

Fig. II.39 ▶

[if $X = 0$, then $k = \left(\sum_i \bar{w}_i \right) / T$ as in Eq. (II.60)] and

$$\bar{v}_m = \frac{q}{k} = \frac{M}{\sum_i \bar{w}_i}$$

as in Eq. (II.82).

The above results show that k and $\bar{v}_m$ are functions of the characteristics of the motion of all vehicles in the observed platoon, whereas q is determined only by the characteristics of the motion of the first and the last vehicle.

In contrast to the observation over a pre-selected distance, as illustrated in Fig. II.38, the platoon can also be observed during a pre-selected time interval T, as illustrated in Fig. II.39. In this case

$$A = XT - \frac{T^2}{2} (\bar{v}_0 + \bar{v}_N)$$

and therefore

$$q = \frac{\sum_i \Delta X_i}{A} = \frac{\sum_i T \bar{v}_i}{XT - \dfrac{T^2}{2} (\bar{v}_0 - \bar{v}_N)} = \frac{\sum_i \bar{v}_i}{X - \dfrac{T}{2} (\bar{v}_0 + \bar{v}_N)}$$

[If $T = 0$, then $q = \left(\sum_i v_i \right) / X$ as in Eq. (II.71).] Also

$$k = \frac{\sum_i \Delta T_i}{A} = \frac{NT}{XT - \dfrac{T^2}{2} (\bar{v}_0 - \bar{v}_N)} = \frac{N}{X - \dfrac{T}{2} (\bar{v}_0 - \bar{v}_N)}$$

[if $T = 0$, then $k = N/X$, as in Eq. (II.70)] and

$$\bar{v}_m = \left(\sum_i \bar{v}_i \right) / N$$

as in Eq. (II.85). Thus k depends only on the characteristics of motion of the first and last vehicle.

II.2.6.2.6 Journey Time Measurement

For the quasi-instantaneous measurement (see Sect. II.2.5.2.3) the measuring distance over which the travel time was measured was comparatively short. If this distance is increased, so that, for example, the travel time between two nodes is measured, then a travel time, rather than a quasi-instantaneous measurement, is referred to.

Journey time measurements are usually made by recording licence plate numbers at the ends of the observed link. It is, however, possible to determine average journey times (and traffic flows) by other observation methods.

II.2.6.2.6.1 Observations Made from a Moving Vehicle

Let us suppose that an observation vehicle travels at v_0 along link L with the stream and counts the active and passive overtakings. Simultaneously, the travel time $r_0 = L/v_0$ is measured.

From Eq. (II.33) we can express the number of active overtakings, M_p^a, on link L as

$$M_p^a(v_0) = k \cdot r_0 \int_0^{v_0} (v_0 - v) \cdot dG_m(v) \tag{II.95}$$

or, taking $v_0 = \dfrac{L}{r_0}$, $v = \dfrac{L}{r}$ and $r \to \infty$ for $v = 0$

$$M_p^a(r_0) = k \cdot r_0 \cdot \int_{r_0}^{\infty} \left(\frac{L}{r_0} - \frac{L}{r} \right) dG_m(r) = k \cdot \frac{L}{r} \int_{r_0}^{\infty} (r - r_0) \cdot dG_m(r)$$

$$= k \cdot v \cdot \int_{r_0}^{\infty} (r - r_0) \cdot dG_m(r) = q \cdot \int_{r_0}^{\infty} (r - r_0) \cdot dG_m(r) \tag{II.96}$$

where $G_m(r)$ is the journey time distribution corresponding to $G_m(v)$. Similarly, from Eq. (II.34) we obtain

$$M_p^p(v_0) = k \cdot r_0 \int_{v_0}^{\infty} (v - v_0) \cdot dG_m(v) \tag{II.97}$$

$$M_p^p(r_0) = q \cdot \int_0^{r_0} (r_0 - r) \cdot dG_m(r). \tag{II.98}$$

Subtracting, we obtain

$$U = M_p^p(r_0) - M_p^a(r_0) = q \left[\int_0^{r_0} (r_0 - r) dG_m(r) - \int_{r_0}^{\infty} (r - r_0) \cdot dG_m(r) \right].$$

Using

$$\int_{r_0}^{\infty} dG_m(r) + \int_0^{r_0} dG_m(r) = 1$$

and

$$\int_{r_0}^{\infty} r \cdot dG_m(r) + \int_0^{r_0} r \cdot dG_m(r) = E_m(r) \, (= \bar{r})$$

we obtain

$$U = q(r_0 - \bar{r}) \tag{II.99}$$

$$\bar{r} = r_0 - \frac{U}{q}. \tag{II.100}$$

(If $U = 0$ then $r_0 = \bar{r}$; for comparison see the "floating car"-method in Sect. II.3.1.2.)

Let us suppose that a second observation vehicle travels with speed v_a along link L against the stream $(r_a = L/v_a)$.

According to Eq. (II.39), the vehicle observes

$$M_b(v_a) = B(v_a) \cdot r_a = k(v_a + \bar{v}_m) \cdot r_a. \tag{II.101}$$

Since $\bar{v}_m = L/r$, [see Eq. (II.82)]

$$M_b(r_a) = k \cdot r_a \left(\frac{L}{r_a} + \frac{L}{\bar{r}} \right) = k \cdot L \cdot r_a \left(\frac{1}{r_a} + \frac{1}{\bar{r}} \right)$$

$$= k \cdot \frac{L}{\bar{r}} (\bar{r} + r_a) = k \cdot \bar{v}_m (\bar{r} + r_a) = q(\bar{r} + r_a) \tag{II.102}$$

From Eqs. (II.99) and (II.102) we obtain

$$q = \frac{U + M_b(r_a)}{r_0 + r_a} \tag{II.103}$$

and substituting in Eq. (II.100)

$$\bar{r} = \frac{M_b(r_a) \cdot r_0 - U \cdot r_a}{U + M_b(r_a)}. \tag{II.104}$$

One could alternatively obtain q and $\bar{r}$ from traffic counts.

II.2.6.2.6.2 Determination of q and $\bar{r}$ from Traffic Counts at the End of a Link

It is assumed that two observation vehicles travel simultaneously along a link of length L with the same constant travel time τ_k, and that no time is lost when turning (Fig. II.40).

Consider a $\tau_k \cdot L$ section of the time-distance plane.

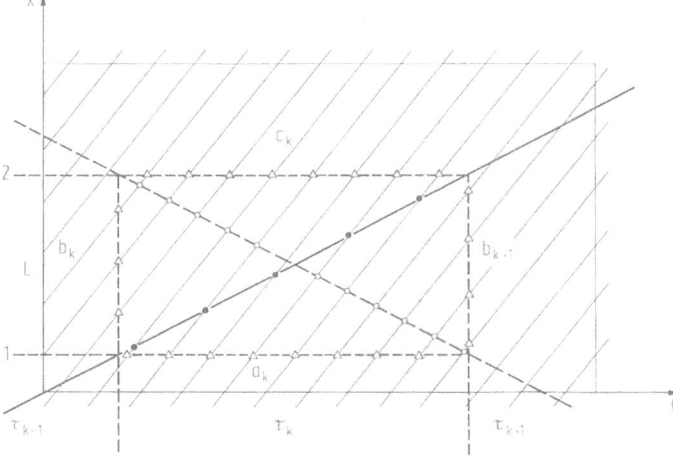

Fig. II.40. (From [46])

Furthermore, let

a_k = the number of vehicles that enter link L in the direction considered
           during interval τ_k

c_k = the number of vehicles that leave link L in the direction considered
           during interval τ_k

b_k = the number of vehicles on link L at the beginning of interval τ_k

b_{k+1} = the number of vehicles on link L at the end of interval τ_k.

Denoting the number of vehicles that overtake the observation vehicle by M_p^p, we see that

$$M_p^p = a_k - b_{k+1} = c_k - b_k.$$

(In general for any line of motion we have that $a_k - b_{k+1} = M_p^p - M_p^a = U$.)
The following number of vehicles

$$N = c_k + b_{k+1} = a_k + b_k = M_b(r_a)$$

encounter the observation vehicle travelling in the opposite direction.

Since it is assumed that both observation vehicles are travelling at the same speed, the following relationships obtain for Eqs. (II.103) and (II.104):

$$r_0 = r_a = \tau$$

and thus

$$q = \frac{U+N}{2 \cdot \tau}$$

and

$$\bar{r} = \frac{(N-U) \cdot \tau}{U+N}.$$

Hence

$$U+N = (a_k - b_{k+1}) + (c_k + b_{k+1}) = a_k + c_k$$

$$N-U = (a_k + b_k) - (a_k - b_{k+1}) = b_k + b_{k+1}.$$

Therefore, for n successive measurement trips each of duration t,

$$q = \frac{\sum\limits_{k=1}^{n} (a_k + c_k)}{2 \cdot n \cdot \tau} \tag{II.105}$$

$$\bar{r} = \frac{\tau \cdot \left(b_1 + b_{n+1} + 2 \cdot \sum\limits_{k+2}^{n} b_k \right)}{\sum\limits_{k=1}^{n} (a_k + c_k)}; \tag{II.106}$$

Since

$$b_{k+1} = a_k + b_k - c_k,$$

both q and $\bar{r}$ may be determined when measurements are made at both ends of the link in successive time intervals of $\bar{r}$, if b_1 is known.

By a simple procedure carried out before the commencement of observations it is possible to determine b_1. Two observers are located together at point 1 (see Fig. II.40). When observer 2 begins to travel from point 1 to point 2, observer 1 commences counting. Observer 2 counts M_p^p or $M_p^p - M_p^a = U$ overtaking manoeuvers. Upon arrival at point 2, observer 2 gives observer 1 a signal, by which time observer 1 has counted a_0 vehicles.

Since M_p^p or $U = a_0 - b_1$ we have

$$a_0 - U = a_0 - a_0 + b_1 = b_1.$$

As from the signal both observers begin counting a_i and c_i respectively for common, equal time intervals τ.

II.2.6.3 Empirical Relationships

II.2.6.3.1 Speed and Volume

Freely flowing traffic is defined as a traffic flow in which each vehicle travels at the desired speed of the driver, without being affected by other vehicles, and subject only to those constraints associated with the vehicle and road characteristics. This kind of flow is imaginable only if very few vehicles are on the road and there are sufficient lanes to allow overtaking without delay in each location at all times. The circular road used repeatedly in the preceding sections as an example assumes such a situation.

Thus the speed of a vehicle in a free flow regime depends only on how fast a driver wishes to travel within the constraints of his vehicle and the road: This speed is called the desired speed. The distribution of the desired speeds depends upon the composition of traffic and upon the road conditions; it is in general a function of distance. It can also be a function of time; as the time of day is known to influence how fast people wish to drive.

The continuing advances in vehicle design have led to a long-term increasing trend in average speed. Figure II.41 illustrates this trend for German and American highways.

The heavier the traffic, the less frequently will vehicles, through a lack of opportunities for overtaking, be able to maintain the desired free speed: Drivers must reduce their speed to that of a slower vehicle more frequently, and for longer times. This results in a continuous decrease in average traffic speed as traffic volume increases. A traffic flow, in which not all vehicles are free to overtake is referred to as partly constrained traffic.

When the lack of overtaking opportunities prevents vehicles from travelling at the desired speed, the result is that these vehicles travel in platoons. A platoon is defined as a line of vehicles in which each vehicle's speed (except the first) is constrained by that of the vehicle ahead.

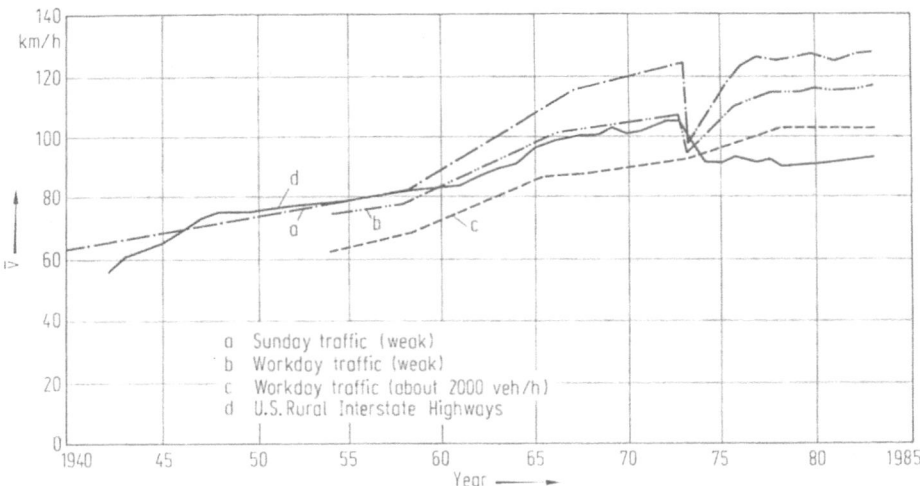

Fig. II.41. (From Dilling, Keller, ACM)

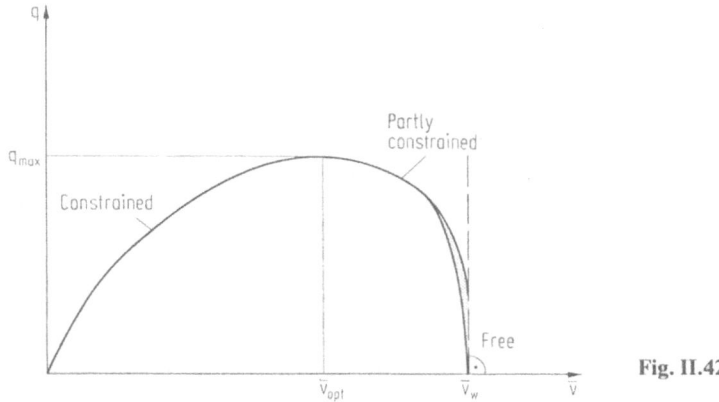

Fig. II.42

The decrease in the average speed begins slowly, even at very small traffic volumes. In practice the free flow regime is defined to include those situations in which the decrease in the average speed can, to a good approximation, be neglected. Figure II.42 illustrates this idea.

In any case, as $\bar{v} \to \bar{v}_w$ ($\bar{v}_w$ = average desired speed)

$$\lim_{\bar{v} \to \bar{v}_w} \frac{d\bar{v}}{dq} \to 0. \tag{II.107}$$

When it is not possible for drivers to carry out their desired overtaking manoeuvres, all vehicles travel in one or more platoons. Several slow drivers can break up the column of traffic into several platoons; the slower drivers do not want to overtake, and the other drivers cannot overtake. This is defined as constrained traffic. The transition from partly constrained to constrained traffic is assumed to

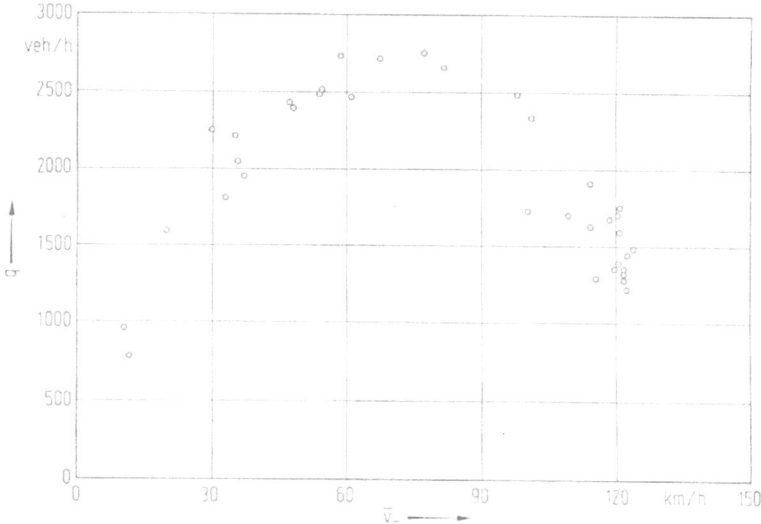

Fig. II.43

occur in the region of the maximum of the curve shown in Fig. II. 42, that is, at the point where

$$\frac{dq}{d\bar{v}} = 0. \tag{II.108}$$

The mean speed at this point is denoted as $\bar{v}_{opt}$ although it is an open question whether traffic flow at the maximum volume is in every respect optimal. As the traffic density increases further (see Sect. II.2.6.3.2) so the average speed decreases to such an extent that the traffic volume decreases. When all vehicles come to a stop, then, by definition, $q = 0$. Figure II.43 shows the relationship between q and $\bar{v}_m$ resulting from observation in one-minute intervals at a construction site on the Cologne-Frankfurt-Autobahn.

The stochastic nature of the traffic flow leads to fluctuations in the measured data; the sketch in Fig. II.42 is therefore only a generalization. Such measurements are easy to make, but fluctuations in the data make them unsuitable for determining the maximum value.

Instead of expressing q as a function of v (or rather, v as a function of q), one can express the slowness w or the journey time r as a function of q. Figure II.44 shows such a relationship corresponding to Fig. II.42.

Such functions are referred to as cost-flow functions and are used for example in transportation planning to assign a traffic stream from a zone i to a zone j to a number of alternative routes according to the traffic flows.

II.2.6.3.2 Speed and Density

The behaviour of a driver depends strongly on how many vehicles he sees on the road (mainly in front of him) and particularly on his distance from the vehicle immediately in front. If a_i are the distance-headways between successive vehicles then

$$k = 1/\bar{a}. \tag{II.109}$$

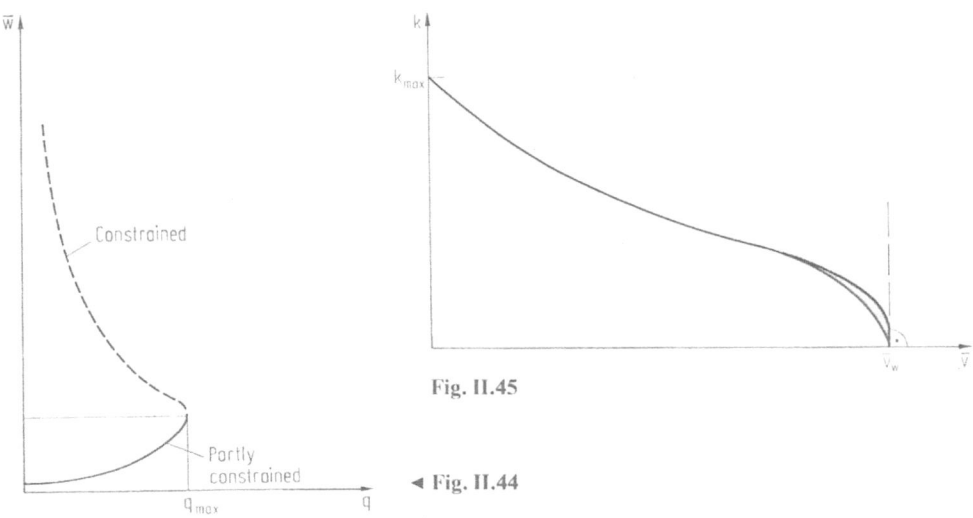

Fig. II.45

◄ Fig. II.44

Observations show that the average speed in the regimes of partly constrained and constrained traffic decreases with increasing traffic density (Fig. II. 45).

In the free flow regime where the mean speed is independent (or nearly independent) of the traffic volume, the mean speed is also independent of the traffic density. In any case, as $k \to 0$,

$$\lim_{k \to 0} \frac{d\bar{v}}{dk} \to 0. \qquad (\text{II}.110)$$

If $\bar{v} = 0$ then $k = k_{max}$. The maximum value of traffic density depends upon the vehicle lengths and on how closely they space themselves when they come to a stop. For road traffic the figure $k_{max} = 150$ veh/km/lane is a rough guideline.

Figure II.46 shows the relationship between k and $\bar{v}_m$ for the same observations used in Fig. II. 43. The data appear to have smaller fluctuations than in Fig. II. 43. For this reason the fundamental diagram is customarily based on the relationship between k and $\bar{v}$, even when the measurement technique measures q_i, and $k_i = q_i \bar{w}_{1_i}$ must be calculated.

Empirical relationships between k and $\bar{v}$ do not clearly mark the location of $\bar{v}_{opt}$. However, if it is assumed that different relationships hold for constrained flow than for partly constrained flow, then presumably $\bar{v}_{opt}$ is located in the region of the assumed discontinuity (Fig. II.47). This question is not yet sufficiently clarified.

II.2.6.3.3 Volume and Density: The Fundamental Diagram

The graphical display of the relationship between volume and density is called the fundamental diagram. Since $\bar{v}_{m_i} = q_i/k_i$ is determined from the slope of a radius vector to a point (q_i, k_i) (Fig. II.48), the fundamental diagram illustrates the relationships among all three parameters q, k, and $\bar{v}_m$.

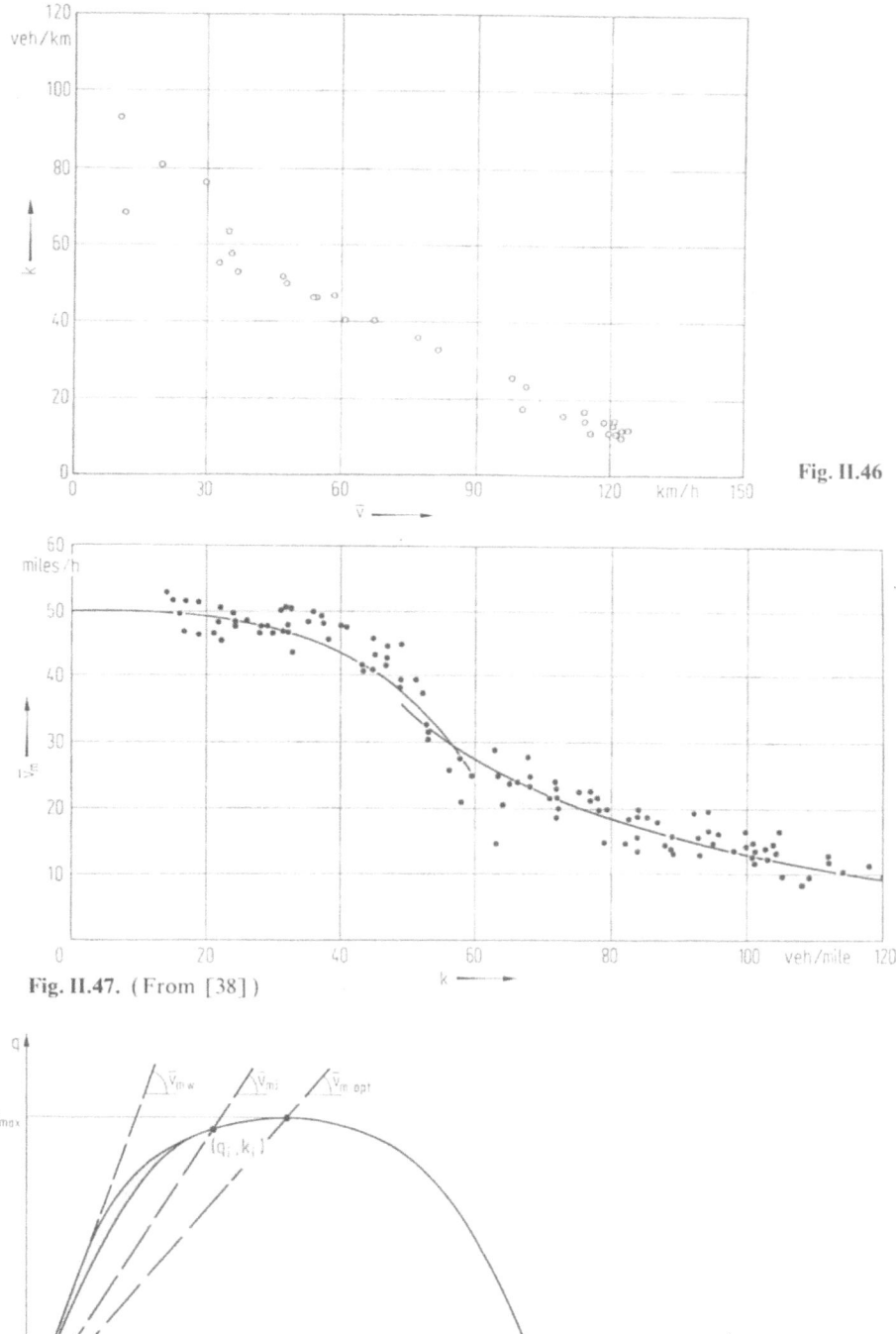

Fig. II.46

Fig. II.47. (From [38])

Fig. II.48

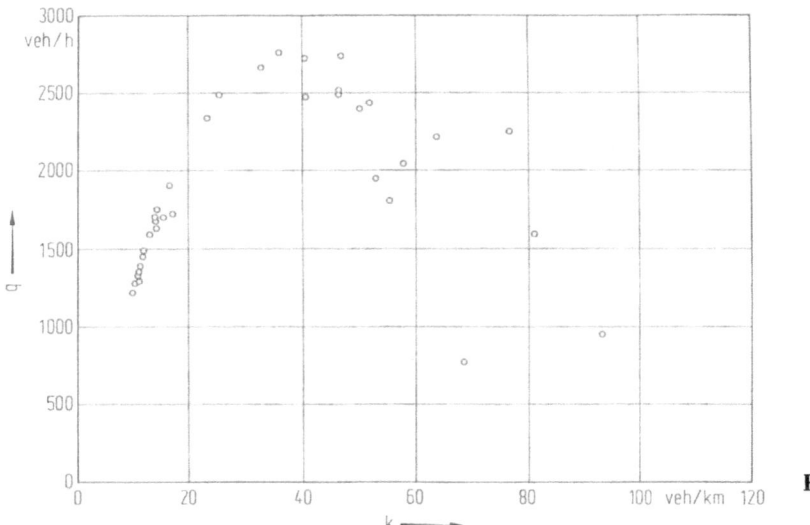

Fig. II.49

In the free-flow regime the fundamental diagram follows approximately the radius vector $\bar{v}_{m_w}$ which is tangential to the fundamental diagram at the origin. Further, the slope of the radius vector associated with q_{max} is $\bar{v}_{m_{opt}}$.

Figure II.49 shows the fundamental diagram for the same measurements shown in Figs. II.43 and II.46. The data exhibit large fluctuations near to the presumed location of the maximum volume, and even larger fluctuation at higher densities.

However, since it is difficult to determine q_{max} directly from a set of measurements, an alternative approach which can often be recommended is to construct the fundamental diagram from a relationship betwen k and $\bar{v}_m$ (see Sect. II.2.6.3.2): Using graphical or other means, one or more smooth curves are drawn through the cloud of data points, as for example in Fig. II.47. The equation $q = k \cdot \bar{v}_m$ is then used to construct the fundamental diagram point by point: for each value k_i, q_i is equal to the area of the rectangle determined by the point (k_i, $\bar{v}_{m_i}$) (Fig. II.50). A fundamental diagram must also satisfy the following boundary conditons:

1. $q = 0$ for $k = 0$ (II.111)

2. $q = 0$ for $k = k_{max}$

3. $\bar{v}_m = \bar{v}_w$ for $k = 0$

4. $\bar{v}_m = 0$ for $k = k_{max}$

5. $\lim\limits_{k \to 0} \dfrac{d\bar{v}_m}{dk} = 0$ or $\lim\limits_{k \to 0} \dfrac{dq}{dk} = \bar{v}_{m_w}$

6. $\dfrac{dq}{dk} = 0$ for $q = q_{max}$.

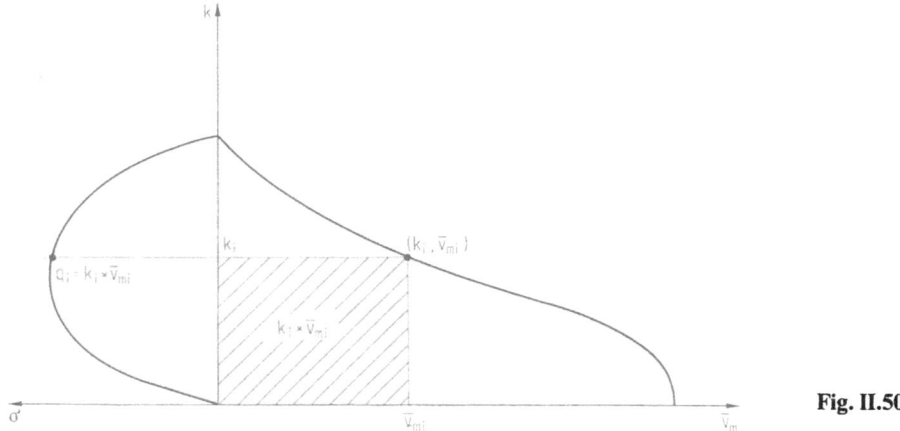

Fig. II.50

In later sections relationships for q as a function of k will be derived from theoretical models. How well such relationships describe traffic flow can be verified using the above boundary conditions.

II.2.6.3.4 The Dynamic Fundamental Diagram

When the measurements over time are plotted in the form of a moving average one obtains what is referred to as the dynamic fundamental diagram. It illustrates particularly clearly how a queue forms at a bottleneck and then disperses behind it.

Figure II.51 shows 50 consecutive values of a 10-minute moving average which is composed of 1-minute observations. The values illustrate the growth of traffic associated with a slight reduction in average speed. From the 17th interval on, the influence of the bottleneck is noticeable. At the bottleneck capacity is exceeded and the resulting queue builds up past the observation (for comparison, see Sect. II.3.3.3.2). Between the 17th and the 23rd intervals the transition from free or partly constrained flow to constrained flow is evident, while between the 24th and the 42nd intervals, constrained flow with a low speed predominates at the point of observation. The queue dissolves between the 43rd and 50th intervals, since the rate of vehicle arrivals has in the meantime decreased.

II.2.6.3.5 Influences on the Shape of the Fundamental Diagram

The shape of the fundamental diagram depends on the conditions under which it was observed. One important factor is the length of the time interval over which the data is aggregated.

Figure II.52 shows the fundamental diagrams obtained from the same series of observations relating to the three-lane carriageway of an autobahn when the interval of measurement is respectively 1, 2, 5 and 10 minutes. It is clear that the distribution of the observation points, particularly in the region of partly constraind and constrained flow, decreases when the length of the measurement interval increases. Moreover, one can better identify the maximum traffic flow that would result from fitting a smooth, continuous function to the data. The

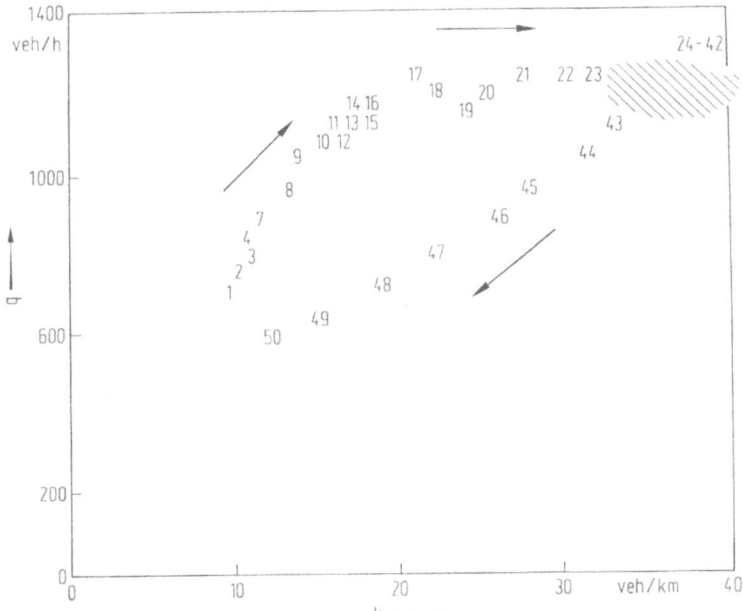

Fig. II.51. (From [44])

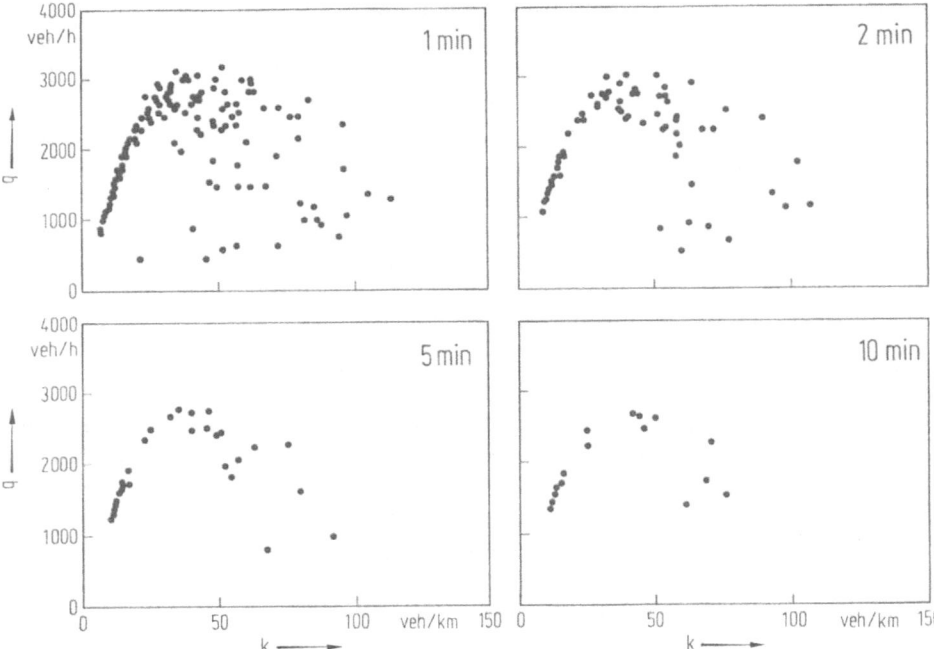

Fig. II.52

shorter the interval of measurement, the more marked the impact of individual slow vehicles and the stochastic element of traffic flow (see Sects. II.2.6.3.4 and II.3.3.1.3).

In addition, there will be clear differences between the fundamental diagrams

- for the individual lanes of a carriageway,
- for the two directions of a two-lane road with opposing traffic,
- with and without speed restrictions,
- before and in a bottleneck,
- under different weather conditions.

In order to compare two fundamental diagrams, it is first necessary to examine carefully the conditions under which the two sets of data were collected in order to ensure comparability.

II.3 Description of the States of Traffic

II.3.1 Freely Flowing Traffic

II.3.1.1 Parameters as Functions of Time and Distance

Freely flowing traffic is defind as traffic in which each driver can travel as he desires completely uninfluenced by the presence of other vehicles (see Sect. II.2.6.3.1). It follows then that the traffic processes in time and distance are independent of each other. If the intensity $\lambda(x,t)$ at the location x is dependent of time, so is the concentration. If the concentration is independent of time, then

$$\frac{\delta \varkappa(x,t)}{\delta t} = 0 \tag{II.112}$$

Differentiating Eq. (II.58) with respect to time we obtain

$$\frac{\delta \varkappa(x,t)}{\delta t} = \frac{\delta \lambda(x,t)}{\delta t} E_1[W(x,t)] + \frac{\delta E_1[W(x,t)]}{\delta t} \lambda(x,t) \tag{II.113}$$

The right-hand side of the equation can only be zero, when there is a functional dependence between the intensity and the mean speed or slowness (but this contradicts the definition of freely flowing traffic (see also Sect. II.2.6.3.1), or when the intensity and the mean speed or slowness are both independent of time. In this fashion a distance-dependent concentration follows from a distance-dependent intensity. When $\lambda(x,t)$ and $\varkappa(x,t)$ are dependent on time and distance, the traffic flow is defined as being non-stationary over time and distance.

In the special case of the circular road, the concentration and the speeds were independent of time, so that the intensity was also independent. If the concentration is independent of time, then must the intensity be independent of distance. This results from the conservation equation to be derived in Sect. II.3.3.3.1.

$$\frac{\delta \varkappa(x,t)}{\delta t} + \frac{\delta \lambda(x,t)}{\delta x} = 0$$

The concentration and speeds for the circular road are likewise independent of distance. In such a case of pure stationarity (stationarity over time and distance) we have

$$\lambda(x,t) = \lambda = \text{const}$$

$$\varkappa(x,t) = \varkappa = \text{const}$$

$$v(x,t) = v = \text{const}.$$

When the speed depends on distance and the intensity is independent of distance because

$$\frac{d\lambda(x)}{dx} = \frac{d\varkappa(x)}{dx} E_m[V(x)] + \frac{dE_m[V(x)]}{dx} \varkappa(x) = 0$$

the concentration must depend on distance. We then have only stationarity over time, for which:

$$\lambda(x,t) = \lambda = \text{const}$$

$$\varkappa(x,t) = \varkappa(x)$$

$$v(x,t) = v(x).$$

If the traffic flow is only stationary over distance, then,

$$\lambda(x,t) = \lambda(t)$$

$$\varkappa(x,t) = \varkappa(t)$$

$$v(x,t) = v(t).$$

As a result of the conservation equation to be derived in Sect. II.3.3.3.1, the concentration is also independent of time:

$$\varkappa(t) = \varkappa = \text{const}.$$

Equations (II.112) and (II.113) require then that the intensity and the slowness (or the speed) must also be independent of time (see above). In freely flowing traffic, stationarity over distance is therefore equivalent to pure stationarity.

Specific relationships connect the parameters which describe freely flowing traffic at different points in the x-t plane. First, it is possible to calculate the expected value of the number of vehicles which can be found in an interval Δx knowing the distribution of travel times over the interval $(x, x + \Delta x)$ and the intensity at location x: A vehicle, which is located at point x at time $t - r$ with speed $v = \Delta x/r = \text{const}$ will be located at $x + \Delta x$ at time t (Fig. II. 53). A vehicle with speed $v_i < v$ and therefore with a travel time $r_i = \Delta x/v_i > r$, located at x at time $t - r$ will still be within the interval Δx at time t. [If the vehicle is not travelling at constant speed, then the same concepts are still valid, if the journey speed $\bar{v}_t$, or $\hat{v}_t$, is used (see Sect. I.2.1).]

The probability that a vehicle arrives in the interval $(t - r, t - r + dr)$ is $\lambda_x(t - r)dr$. Let $f(r|x,t)$ be the probability function of the travel times $r(x,t)$, and

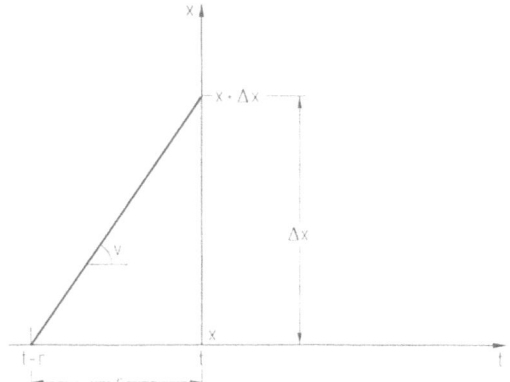

Fig. II.53

$F(r|x,t)$ the distribution function. Then, the probability that a vehicle needs a travel time $(r, r+dr)$ to traverse the distance Δx is

$$P(r<R<r+dr|x,t) = f(r|x,t)\,dr = dF(r|x,t)$$

and the probability that a vehicle requires a time $r(x,t)>r$ is

$$\int_r^\infty f(r|x,t)\,dr = \int_r^\infty dF(r|x,t) = 1 - \int_0^r dF(r|x,t) = 1 - F(r|x,t).$$

The probability that a vehicle arrives in the time interval $(t-r,\ t-r+dr)$ and requires a travel time $r(x,t)>r$ to traverse Δx is

$$[\lambda_x(t-r)\,dr]\,[1-F(r|x,t-r)]. \tag{II.114a}$$

Considering not just the single time interval $(t-r, t-r+dr)$, but all possible time intervals, integration of Eq.(II.114a) over all values of r yields the expected number of vehicles in Δx:

$$E[N(t,x,\Delta x)] = \int_0^\infty [1-F(r|x,t-r)]\lambda_x(t-r)\,dr. \tag{II.114b}$$

If the traffic flow is independent of time $(\lambda_x = \text{const})$ then

$$E[N(x,\Delta x)] = \lambda_x \int_0^\infty [1-F(r|x)]\,dr.$$

Since the expected value of the random value r is defined as

$$E(R) = \int_0^\infty r\,f(r|x)\,dr$$

and since an integration by parts gives the relationship

$$E(R) = \int_0^\infty r\,f(r|x)\,dr = \int_0^\infty [1-F(r|x)]\,dr$$

we can now write

$$E[N(x,\Delta x)] = \lambda_x E(R). \tag{II.115}$$

The result, which includes the distance travelled during the fixed time interval Δt, can be similarly derived.

$$E[M(x,t,\Delta t)] = \int_0^\infty [1 - F(s|t, x-s)] \varkappa_t(x-s) ds \qquad (II.116)$$

When we have stationarity over distance, the expected number of vehicles during time interval Δt is given as

$$E[M(t,\Delta t)] = \varkappa_t E(S). \qquad (II.117)$$

Notice the similarity of Eqs. (II.115) and (II.117) with Eqs. (II.57) and (II.58). When the value of the parameter λ at a location x is known, then the knowledge of either the speed distribution or the journey time distribution permits the calculation of the value of this parameter at the location $x + \Delta x$; this same statement applies also to the calculation of the value of $\varkappa$ at time $t + \Delta t$ based on its value at time t. Let $g_1(v) dv = dG_1(v)$ be the probability that a vehicle which appears at location x has a speed $(v, v + dv)$. Then the probability, that this vehicle also arrives at $(t, t + dt)$ is

$$\lambda_x(t) dt\, dG_1(v).$$

If this vehicle is assumed to travel at constant speed then $r = \Delta x/v$, and

$$\lambda_{x+\Delta x}(t) = \int_0^\infty \lambda_x\left(t - \frac{\Delta x}{v}\right) dG_1\left(v, t - \frac{\Delta x}{v}\right).$$

In corresponding fashion one also obtains

$$\varkappa_{t+\Delta t}(x) = \int_0^\infty \varkappa_t(x - vt) dG_m(v, x - vt).$$

If, instead of the above situation, we now insert travel time, r, and travel distance, s, in order to be able to characterize any arbitrary travel path, we then obtain

$$\lambda_{x+\Delta x}(t) = \int_0^\infty \lambda_x(t-r) dF(r|x, t-r)$$

or

$$\varkappa_{t+\Delta t}(x) = \int_0^\infty \varkappa_t(x-s) dF(s|t, x-s).$$

When the travel times are time-independent the first equation above becomes

$$\lambda_{x+\Delta x}(t) = \int_0^\infty \lambda_x(t-r) dF(r|x).$$

When the travel distance is distance-independent the second equation above becomes

$$\varkappa_{t+\Delta t}(x) = \int_0^\infty \varkappa_t(x-s) dF(s|t).$$

II.3.1.2 Overtaking in Freely Flowing Traffic

By definition freely flowing traffic allows unhindered overtaking. If we know the distribution of travel time per distance interval $F(r|x,\Delta x,t)$ or the distribution of travel distance per time interval $F(s|t,\Delta t,x)$, then we can calculate the expected number of overtakings per distance interval Δx or per time interval Δt (see Fig. II.54).

Let us assume that within the considered segment of the x-t plane all trajectories are straight lines and, therefore, any two trajectories can have at most one intersection. The number of overtakings for a vehicle with travel time r_0 is equal to the number of intersections of its trajectory with trajectories of other vehicles. A vehicle, which arrives at location x at time t_0 and requires travel time r_0 to traverse the distance interval $(x,x+\Delta x)$, overtakes during r_0 those slower vehicles (i.e. for which $r > r_0$), which have previously passed the point x during the time interval $(t_0 - r + r_0, t_0)$ (see Fig. II.54). The expected value of this number is

$$\int_{t_0-r+r_0}^{t_0} \lambda_x(t)\,dt\,dF(r|x,\Delta x,t)\ \text{veh}.$$

To calculate the number of overtakings in the general case of non-stationary traffic results in a very unpleasant formula. Therefore, the intensity and the travel time distribution $F(r|x,\Delta x,t)$ will be assumed to be independent of time in the following derivation (see Sect. II.3.1.1). The number of active overtakings carried out by a vehicle having travel time r_0 is then

$$\int_{t_0-r+r_0}^{t_0} \lambda\,dF(r|x,\Delta x)\,dt = \lambda\,dF(r|x,\Delta x)\int_{t_0-r+r_0}^{t_0} dt = \lambda\,dF(r|x,\Delta x)\,(r-r_0)$$

$$(\text{II}.118)$$

and the number of passive overtakings for this same vehicle is

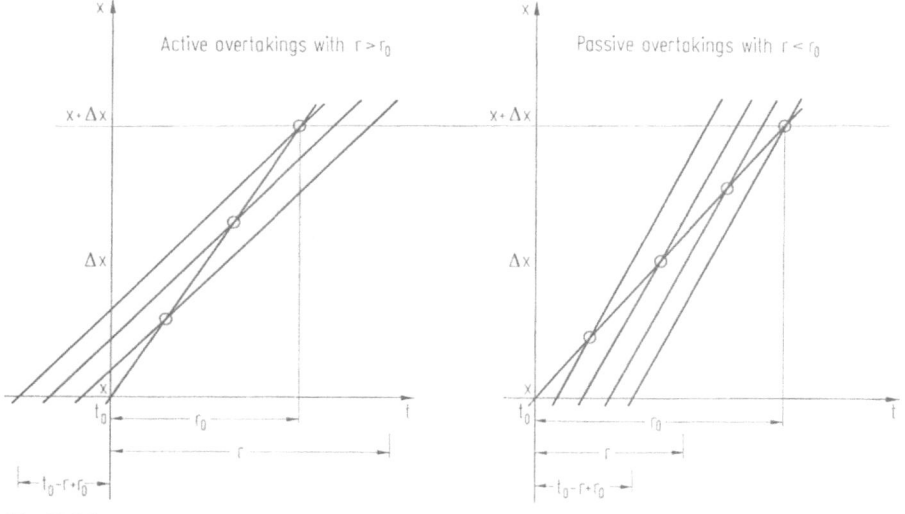

Fig. II.54

$$\int_{t_0}^{t_0-r+r_0} \lambda \, dF(r|x,\Delta x) \, dt = \lambda \, dF(r|x,\Delta x)(r_0-r) \tag{II.119}$$

(see Fig. II.54). Integrating the two preceding equations over r, the expected total number of active and passive overtakings, respectively, in the interval Δx is obtained as

$$E[M_p^a(r_0)] = \lambda \int_{r_0}^{\infty} (r-r_0) \, dF(r|x,\Delta x) \tag{II.120}$$

and

$$E[M_p^p(r_0)] = \lambda \int_0^{r_0} (r_0-r) \, dF(r|x,\Delta x). \tag{II.121}$$

The slowness w is defined as the required time per distance unit (see Sect. I.1.4). Therefore, the distribution of the travel time over a unit distance is identical with the distribution of the slowness w. Dividing Eqs. (II.120) and (II.121) by Δx we therefore obtain the expected number of active and passive overtakings, respectively, per unit distance, for a vehicle with slowness w_0 as

$$E[k_p^a(w_0)] = \lambda \int_{w_0}^{\infty} (w-w_0) \, dG_1(w) \tag{II.122}$$

and

$$E[k_p^p(w_0)] = \lambda \int_0^{w_0} (w_0-w) \, dG_1(w). \tag{II.123}$$

We will name these measures overtaking density. Analogously, the number of active and passive overtakings, respectively, per unit time can be calculated for a vehicle having v_0 using the distribution of travel distance per time interval Δt as

$$E[q_p^a(v_0)] = \varkappa \int_0^{v_0} (v_0-v) \, dG_m(v) \tag{II.124}$$

and

$$E[q_p^p(v_0)] = \varkappa \int_{v_0}^{\infty} (v-v_0) \, dG_m(v). \tag{II.125}$$

These equations are identical with Eqs. (II.33) and (II.34) in Sect. II.2.3.3. Let $M_p^{a+p}(v_0)$ be the number of overtaking manoeuvers that a vehicle driving at v_0 makes in a finite time interval Δt. Then

$$q_p^{a+p}(v_0) = M_p^{a+p}(v_0)/\Delta t$$

and consequently $M_p^{a+p}(v_0) = q_p^{a+p}(v_0) \cdot \Delta t$.

Since $\Delta t = \Delta x/v_0$ and replacing $\varkappa$ with k and λ with q

— the number of active overtaking manoeuvers made by a vehicle travelling at v_0 along a link of length Δx is

$$M_p^a(v_0,\Delta x) = \frac{k \cdot \Delta x}{v_0} \int_0^{v_0} (v_0-v) \, dG_m(v) \tag{II.126}$$

- the number of passive overtaking manoeuvers made by a vehicle travelling at v_0 along a link of length Δx is

$$M_p^p(v_0,\Delta x) = \frac{k \cdot \Delta x}{v_0} \int\limits_{v_0}^{\infty} (v - v_0) dG_m(v) \qquad (II.127)$$

- the total number of overtaking manoeuvers made by a vehicle travelling at v_0 is

$$M_p^{a+p}(v_0,\Delta x) = \frac{k \cdot \Delta x}{v_0} \left[E_m(V) - v_0 + 2 \int\limits_0^{v_0} (v_0 - v) dG_m(v) \right] \qquad (II.128)$$

[compare with Eq. (II.35)].

Usually q and $g_l(v)$ rather then k and $g_m(v)$ are known. Since $q = k \cdot \bar{v}_m$ [see Eq. (II.25)] and $dG_m(v)/\bar{v}_m = dG_l(v)/v$ [see Eq. (II.26)] we obtain

$$M_p^a(v_0,\Delta x) = \frac{q \cdot \Delta x}{v_0} \int\limits_0^{v_0} \left(\frac{v_0}{v} - 1 \right) dG_l(v) \qquad (II.129)$$

$$M_p^p(v_0,\Delta x) = \frac{q \cdot \Delta x}{v_0} \int\limits_{v_0}^{\infty} \left(1 - \frac{v_0}{v} \right) dG_l(v). \qquad (II.130)$$

Noting that $dG_l(w) = dG_l(1/v)$ and taking $w = 1/v$ we obtain from Eq. (II.129) the following

$$M_p^a(v_0,\Delta x) = \frac{q \cdot \Delta x}{v_0} \left[v_0 \cdot \int\limits_0^{v_0} \frac{1}{v} dG_l(v) - \int\limits_0^{v_0} dG_l(v) \right]$$

$$= \frac{q \cdot \Delta x}{v_0} \left[v_0 \cdot \int\limits_{w_0}^{\infty} w\, dG_l(w) - \int\limits_{w_0}^{\infty} dG_l(w) \right] \qquad (II.131)$$

and from Eq. (II.130)

$$M_p^p(v_0,\Delta x) = \frac{q \cdot \Delta x}{v_0} \left[\int\limits_0^{w_0} dG_l(v) - v_0 \int\limits_0^{w_0} w\, dG_l(v) \right]. \qquad (II.132)$$

Therefore, the following relationship holds

$$M_p^{a+p} = \frac{q \cdot \Delta x}{v_0} \left[v_0 \cdot \int\limits_{w_0}^{\infty} w\, dG_l(w) - \int\limits_{w_0}^{\infty} dG_l(w) + \int\limits_0^{w_0} dG_l(w) - v_0 \int\limits_0^{w_0} w\, dG_l(w) \right]$$

$$= \frac{q \cdot \Delta x}{v_0} \left[v_0 \cdot E_l(W) - 1 + 2v_0 \int\limits_0^{w_0} w_0 dG_l(w) - 2v_0 \int\limits_0^{w_0} w\, dG_l(w) \right]$$

$$= \frac{q \cdot \Delta x}{v_0} \left[v_0 \cdot E(W) - 1 + 2v_0 \int\limits_0^{w_0} (w_0 - w) dG_l(w) \right]$$

$$= q \cdot \Delta x \cdot \left[E_l(W) - w_0 + 2 \int\limits_0^{w_0} (w_0 - w) dG_l(w) \right] \qquad (II.133)$$

where $w_0 = 1/v_0$ represents the slowness of the observation vehicle, and $E_l(W)$ the locally measured average slowness.

It should be noted that, in general,

$$E_l(W) = \frac{1}{E_m(V)} \mp \frac{1}{E_l(V)}.$$

Correspondingly, let $M_p^{a+p}(w_0)$ be the number of overtaking manoeuvers over a finite distance Δx. Then

$$k_p^{a+p}(w_0) = \frac{M_p^{a+p}(w_0)}{\Delta x}.$$

Consequently, $M_p^{a+p}(w_0) = k_p^{a+p}(w_0) \cdot \Delta x$, and therefore [from Eq. (II.107)]

— the number of active overtaking manoeuvers of a vehicle travelling with slowness w_0 over distance Δx is

$$M_p^a(w_0) = q \cdot \Delta x \int_{w_0}^{\infty} (w - w_0) dG_l(w)$$

— the number of passive overtaking manoeuvers of a vehicle travelling with slowness w_0 over distance Δx is

$$M_p^p(w_0) = q \cdot \Delta x \int_0^{w_0} (w_0 - w) dG_l(w)$$

— the number of all overtaking manoeuvers of a vehicle travelling with slowness w_0 over distance Δx is

$$M_p^{a+p}(w_0) = q \cdot \Delta x \left[\int_0^{w_0} (w_0 - w) dG_l(w) + \int_{w_0}^{\infty} (w - w_0) dG_l(w) \right].$$

Taking

$$\int_{w_0}^{\infty} w \cdot dG_l(w) = E_l(W) - \int_0^{w_0} w \cdot dG_l(w)$$

and

$$\int_{w_0}^{\infty} dG_l(w) = 1 - \int_0^{w_0} dG_l(w)$$

we obtain

$$M_p^{a+p}(w_0) = q \cdot \Delta x \left[E_l(W) - w_0 + 2 \cdot \int_0^{w_0} (w_0 - w) dG_l(w) \right].$$

Example 34. Consider again the example of the ring road introduced in Sect. II.2.3.2. Let us assume that an observation vehicle travels at $v_0 = 70$ km/h round the road of length $L = 1$ km. The number of active overtaking manoeuvers per lap can be calculated from the instantaneous speed distribution using Eq. (II.126)

$$M_p^a(70,1) = \frac{4 \cdot 1}{70} \sum_{v_i \leq v_0} (v_0 - v_i) dG_m(v_i)$$

$$= \frac{4}{70} (50 \cdot 0.25 + 30 \cdot 0.25 + 10 \cdot 0.25) = \frac{9}{7}$$

Similarly, the number of passive overtaking manoeuvers is obtained via Eq. (II.127)

$$M_p^p(70,1) = \frac{4\cdot 1}{70} \sum_{v_i \geq v_0} (v - v_0) dG_m(v_i) = \frac{4}{70}\cdot(10\cdot 0.25) = \frac{1}{7}$$

The total number of overtaking manoeuvers made by the vehicle is therefore

$$M_p^{a+p} = \frac{9}{7} + \frac{1}{7} = \frac{10}{7}.$$

The same answer could also have been obtained from Eq. (II.128):

$$M_p^{a+p}(70,1) = \frac{4\cdot 1}{70}\left[50 - 70 + 2\sum_{v_i \geq v_0}(v_0 - v_i) dG_m(v_i)\right] = \frac{4}{70}\cdot 25 = \frac{10}{7}.$$

These values could also have been obtained from the local speed distribution

$$M_p^p(70,1) = \frac{200\cdot 1}{70} \sum_{v_i \leq v_0}\left(\frac{v_0}{v} - 1\right) dG_1(v_i)$$

$$= \frac{20}{7}\left[\left(\frac{70}{20} - 1\right)\cdot 0.1 + \left(\frac{70}{40} - 1\right)\cdot 0.2 + \left(\frac{70}{60} - 1\right)\cdot 0.3\right] = \frac{9}{7}$$

$$M_p^p(70,1) = \frac{200\cdot 1}{70} \sum_{v_i \geq v_0}\left(1 - \frac{v_0}{v}\right) dG_1(v_i)$$

$$= \frac{20}{7}\left[\left(1 - \frac{70}{80}\right)\cdot 0.4\right] = \frac{1}{7}.$$

Finally, the total number of overtaking manoeuvers, M_p^{a+p} may be obtained directly from Eq. (II.133):

$$E_1(W) = \frac{1}{20}\cdot 0.1 + \frac{1}{40}\cdot 0.2 + \frac{1}{60}\cdot 0.3 + \frac{1}{80}\cdot 0.4 = \frac{1}{50}$$

and hence:

$$M_p^{a+p} = 200\cdot 1\left[\frac{1}{50} - \frac{1}{70} + 2\sum_{w_i \leq w_0}(w_0 - w_i) dG_1(w_i)\right]$$

$$= 200\left[\frac{1}{50} - \frac{1}{70} + 2\cdot\left(\frac{1}{70} - \frac{1}{80}\right)\cdot 0.4\right] = \frac{10}{7}$$

There is an alternative derivation of Eq. (II.126) to (II.128).

A vehicle P_0 travels at speed v_0 along a link of length L from A to C. The journey time is $\Delta t_0 = L/v_0$. If another vehicle P travels simultaneously at v from B to C (see Fig. II.55), it requires $\Delta t = x/v$. P_0 can only overtake P on link L if $\Delta t_0 < \Delta t$ (see also Fig. II.54):

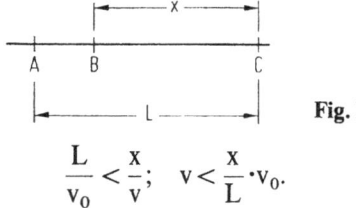

Fig. II.55

$$\frac{L}{v_0} < \frac{x}{v}; \quad v < \frac{x}{L}\cdot v_0.$$

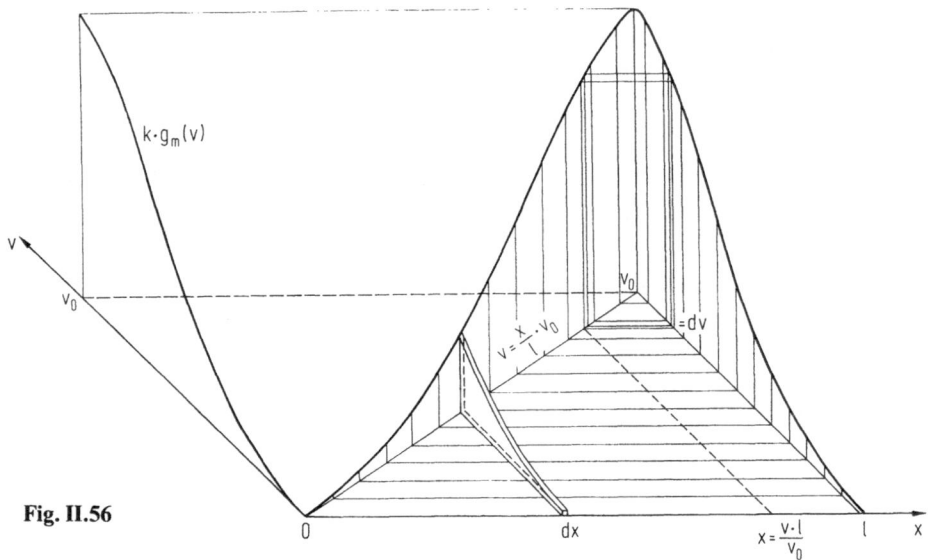

Fig. II.56

On link L there are k vehicles whose speeds are distributed according to $G_m(v)$. On a section of length dx there are $k \cdot dx$ vehicles whose speeds are also distributed according to $G_m(v)$. Among these, only

$$k \cdot dx \cdot \int_0^{x \cdot v_0/L} dG_m(v)$$

satisfy the above inequality. Thus, vehicle P_0 can overtake

$$M_p^a(v_0, \Delta x) = \int_{x=0}^{x=L} k \cdot \left[\int_{v=0}^{v=x \cdot v_0/L} dG_m(v) \right] dx$$

vehicles on link L.

The calculation of the above double integral corresponds to the calculation of a volume. As shown by Fig. II. 56, the frequency distribution $g_m(v)$ forms a body with a roof-shaped surface in the x-v plane. Of this, the volume cut off by the vertical plane through the line

$$v = x \cdot v_0/L$$

corresponds to the number of active overtaking manoeuvers. The calculation of this volume may be based on the summation of infinitely thin slices lying parallel to either the v-axis or the x-axis:

$$M_p^a(v_0, \Delta x) = k \cdot \int_{v=0}^{v=v_0} \left[\int_{x=v \cdot L/v_0}^{x=L} g_m(v) dx \right] dv = k \cdot \int_{v=0}^{v=v_0} \left(L - \frac{v \cdot L}{v_0} \right) dG_m(v)$$

$$= \frac{k \cdot L}{v_0} \int_{v=0}^{v=v_0} (v_0 - v) dG_m(v).$$

The above equation is identical to Eq. (II.126) when $\Delta x = L$.

In the derivations presented so far, it has been assumed that the distribution of speed and slowness are continuous. One could, however, obtain the number of overtaking manoeuvers from discrete distributions of these magnitudes.

Let us suppose that the traffic stream is composed of components of magnitude Δq_i with respective speeds v_i. For each component

$$\Delta q_i = v_i \cdot \Delta k_i$$

where

$$\sum^i \Delta q_i = q \quad \text{and} \quad \sum^i \Delta k_i = k.$$

If an observation vehicle travels at v_0 with the stream, all vehicles travelling at $v_i < v_0$ will be overtaken (active overtaking) while it will be overtaken by all vehicles travelling at $v_j > v_0$ (passive overtaking). Let the number of active and passive overtaking manoeuvers per interval of time (the overtaking flow) be q_p^a and q_p^p respectively:

$$\Delta q_p^{ai} = \Delta k_i (v_0 - v_i) = \Delta k_i \cdot v_0 - \Delta q_i.$$

On a link of length L we have the relationships

$$t_i = L/v_i, \quad t_j = L/v_j, \quad t_0 = L/v_0, \quad v_i < v_0 < v_j$$

the observation vehicle overtakes the following number of vehicles travelling at v_i:

$$\Delta M_p^{ai} = \Delta q_p^{ai} \cdot t_0 = \Delta k_i (v_0 - v_i) \cdot t_0 = \Delta k_i \left(\frac{L}{t_0} - \frac{L}{t_i} \right) \cdot t_0$$

$$= \Delta k_i \cdot L \left(1 - \frac{t_0}{t_i} \right) = \Delta k_i \cdot \frac{L}{t_i} (t_i - t_0)$$

$$= \Delta k_i \cdot v_i (t_i - t_0) = \Delta q_i (t_i - t_0) \qquad (\text{II.134})$$

and is overtaken passively by the following number of vehicles

$$\Delta M_p^{pj} = \Delta q_j (t_0 - t_j). \qquad (\text{II.135})$$

This may be illustrated graphically.

Figure II.57 shows the movement of a subset of vehicles travelling at $v_j > v_0$. The number of vehicles on the link AB (which is equal to the number of vehicles at time interval T at point A) is $M_1 = \Delta q_j T$, and the number of vehicles on the link 0A

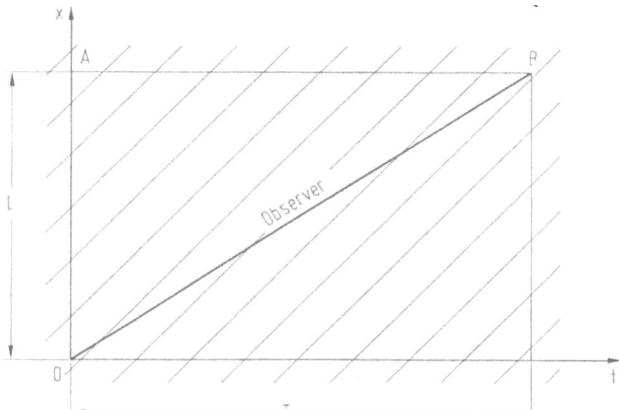

Fig. II.57

at time $t=0$ is $M_2 = \Delta k_j L$. Consequently, the observation vehicle is overtaken by the following number of vehicles:

$$\Delta M_p^{p_j} = \Delta q_j \cdot T - \Delta k_j L. \tag{II.136}$$

When Eq. (II.136) is divided by T, the number of overtaking manoeuvers per interval of time, namely the overtaking flow, is

$$\frac{\Delta M_p^{p_j}}{T} = \Delta q_p^{p_j} = \Delta q_j - \Delta k_j \frac{L}{T}$$

or with

$$L/T = v_0$$
$$= \Delta q_j - \Delta k_j \cdot v_0$$
$$= \Delta k_j (v - v_0). \tag{II.137}$$

Alternatively, dividing Eq. (II.136) by L yields the number of overtaking manoeuvers per interval of distance, or the overtaking density:

$$\frac{\Delta M_p^{p_j}}{L} = \Delta k_p^{p_j} = \Delta q_j \frac{T}{L} - \Delta k_j$$

or with $w_0 = T/L$

$$\Delta k_p^{p_j} = \Delta q_j \cdot w_0 - \Delta k_j. \tag{II.138}$$

Since

$$k = q \cdot w$$

Equation (II.138) may also be written as follows

$$\Delta k_p^{p_j} = \Delta q_j \cdot w_0 - \Delta q_j \cdot w_j = \Delta q_j (w_0 - w_j). \tag{II.139}$$

The total overtaking flow and density for the observation vehicle may be obtained by summation:

$$q_p^a = \sum_i \Delta k_i (v_0 - v_i) = k_1 v_0 - q_1$$

$$q_p^p = \sum_j \Delta k_j (v_j - v_0) = q_2 - k_2 \cdot v_0$$

$$k_p^a = \sum_i \Delta q_i (w_i - w_0) = k_1 - q_1 \cdot w_0$$

$$k_p^p = \sum_j \Delta q_j (w_0 - w_j) = q_2 w_0 - k_2$$

where

$$k_1 = \sum_i \Delta k_i; \quad k_2 = \sum_j \Delta k_j; \quad k_1 + k_2 = k$$

$$q_1 = \sum_i \Delta q_i; \quad q_2 = \sum_j \Delta q_j; \quad q_1 + q_2 = q.$$

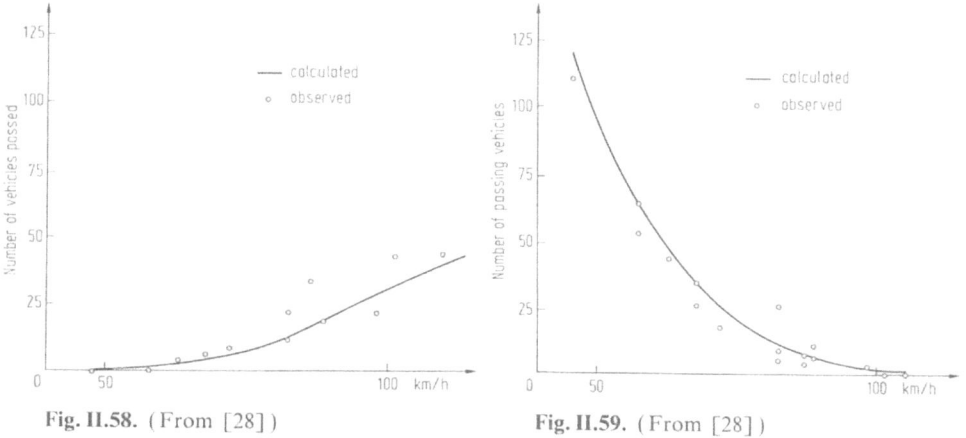

Fig. II.58. (From [28])

Fig. II.59. (From [28])

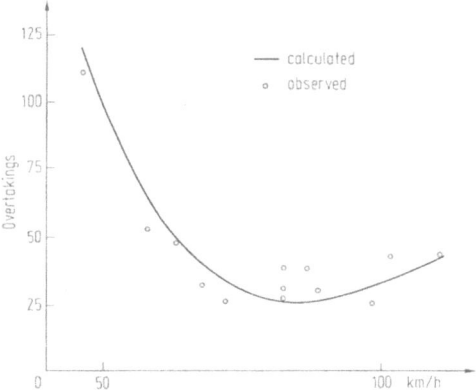

Fig. II.60. (From [28])

In order to test the validity of Eqs. (II.126) to (II.128), measurements were made on a section of an autobahn. During a period of stable speed behaviour, observation vehicles travelling at different speeds, v_0, were fed into the stream of vehicles, and the number of active and passive overtaking manoeuvers were recorded. In Figs. II.58 to II.60 the observed numbers of overtaking manoeuvers in relation to speed are compared with their estimated values. Figure II.58 relates to active overtaking manoeuvers, Fig. II.59 to passive overtaking manoeuvers and Fig. II.60 to the total number of observed overtaking manoeuvers.

From Eqs. (II.126) and (II.127) we obtain

$$M_p^a(v_0, \Delta x) - M_p^p(v_0, \Delta x) = \frac{k \cdot \Delta x}{v_0} \left[\int_0^{v_0} (v_0 - v) dG_m(v) - \int_{v_0}^{\infty} (v - v_0) dG_m(v) \right]$$

$$= \frac{k \cdot \Delta x}{v_0} \cdot \int_0^{\infty} (v_0 - v) dG_m(v).$$

With

$$\int_0^\infty dG_m(v) = 1 \quad \text{and} \quad \int_0^\infty v \cdot dG_m(v) = E_m(V)$$

we arrive at

$$M_p^a(v_0, \Delta x) - M_p^p(v_0, \Delta x) = \frac{k \cdot \Delta x}{v_0} [v_0 - E_m(V)]$$

$$E_m(V) = v_0 \left[1 - \frac{M_p^a(v_0, \Delta x) - M_p^p(v_0, \Delta x)}{k \cdot \Delta x} \right]. \tag{II.140}$$

If a vehicle travels at v_0 so that the number of active and passive overtaking manoeuvers are equal, then

$$v_0 = E_m(v)$$

and the vehicle is travelling at the space-mean speed of the traffic stream. When observations are carried out under these conditions, it is referred to as the "floating car" method.

Dividing Eqs. (II.126) to (II.128) by Δx, yields the number of overtaking manoeuvers made by one vehicle travelling at v_0 per unit distance (for example, per kilometer). However, $q \cdot dG_1(v_0)$ vehicles per unit of time pass a measurement point at v_0. Consequently, again taking $k = q/\bar{v}_m$ and $dG_m = dG_1 \cdot v_m/v$

— the number of active overtaking manoeuvers per unit time and distance of all vehicles travelling at v_0

$$R_p^a(v_0) = \frac{q^2 \cdot dG_1(v_0)}{\bar{v}_m \cdot v_0} \int_0^{v_0} (v_0 - v) dG_m \tag{II.141}$$

$$= \frac{q^2 \cdot dG_1(v_0)}{v_0} \cdot \int_0^{v_0} \left(\frac{v_0}{v} - 1 \right) dG_1(v) \tag{II.142}$$

— the number of passive overtaking manoeuvers per unit time and distance for all vehicles travelling at v_0

$$R_p^p(v_0) = \frac{q^2 \cdot dG_1(v_0)}{\bar{v}_m \cdot v_0} \cdot \int_{v_0}^\infty (v - v_0) \cdot dG_m(v) \tag{II.143}$$

$$= \frac{q^2 \cdot dG_1(v_0)}{v_0} \cdot \int_{v_0}^\infty \left(1 - \frac{v_0}{v} \right) \cdot dG_1(v) \tag{II.144}$$

— and the total number of active or passive overtaking manoeuvers per unit time and distance of the whole stream (referred to as the overtaking rate) is

$$R_p^a = \frac{q^2}{\bar{v}_m} \int_{v_0=0}^{v_0=\infty} \frac{1}{v_0} \left[\int_{v=0}^{v_0} (v_0 - v) dG_m(v) \right] dG_1(v_0) \tag{II.145}$$

$$= q^2 \cdot \int_{v_0=0}^{v_0=\infty} \frac{1}{v_0} \left[\int_{v=0}^{v_0} \left(\frac{v_0}{v} - 1 \right) \cdot dG_1(v) \right] dG_1(v_0) \tag{II.146}$$

$$R_p^p = \frac{q^2}{\bar{v}_m} \cdot \int_{v_0=0}^{v_0=\infty} \frac{1}{v_0} \left[\int_{v=v_0}^{v=\infty} (v-v_0) dG_m(v) \right] dG_1(v_0) \qquad (II.147)$$

$$= q^2 \cdot \int_{v_0=0}^{v_0=\infty} \frac{1}{v_0} \left[\int_{v=v_0}^{v=\infty} \left(1-\frac{v_0}{v}\right) \cdot dG_1(v) \right] dG_1(v_0). \qquad (II.148)$$

In the case of discrete speed distributions, the integrals are replaced by summations. Hence, Eq.(II.145) becomes

$$R_p^a = \frac{q^2}{\bar{v}_m} \sum_{v_0=0}^{v_0=\infty} \frac{1}{v_0} \left[\sum_{v=0}^{v=v_0} (v_0-v) \right]. \qquad (II.149)$$

The special case where a two-lane carriageway has on the inside lane a platoon with parameters (q_1,k_1,v_1) and on the outside lane another platoon with parameters (q_2,k_2,v_2) is now considered. It is assumed that $v_2 > v_1$. The total number of active overtaking manoeuvers per unit time and distance is required. It should be noted that the term q^2 in Eq.(II.145) is the product of the number of vehicles overtaking and the number of vehicles overtaken.

Setting $\bar{v}_m = v = v_1$ and $v_0 = v_2$, Eq.(II.149) becomes

$$R_p^a = \frac{q_1}{v_1} \cdot \frac{q_2}{v_2} \cdot (v_2-v_1).$$

Taking $q_i/v_i = k_i$, we obtain

$$R_p^a = k_1 \cdot k_2 (v_2-v_1).$$

There is an alternative derivation of this equation. Consider a vehicle travelling at speed v_2 and over distance X (see Fig. II.61). There are

$$m_p^a(v_2,X) = q_{0_a} \cdot \Delta t = k_1 (v_2-v_1) \cdot \frac{X}{v_2}$$

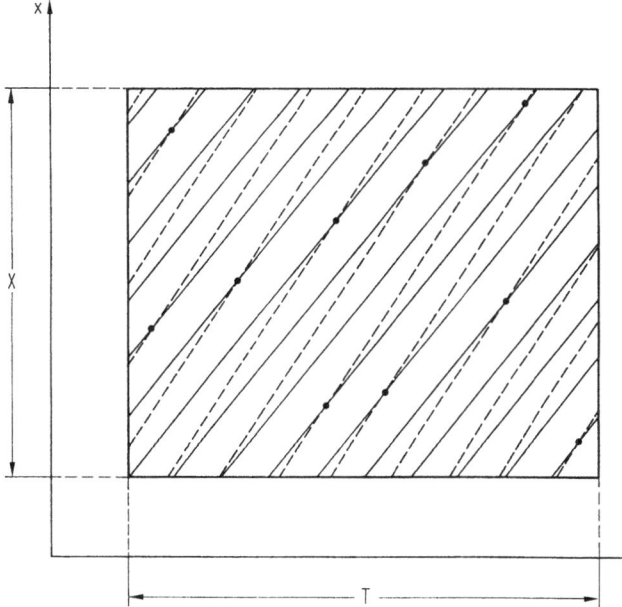

Fig. II.61

vehicles in the lane. During the time interval T, $q_2 \cdot T = k_2 \cdot v_2 \cdot T$ vehicles travelling at v_2 enter the section under consideration. The total number of active overtaking manoeuvers in the region $X \cdot T$ is

$$M_p^a = k_1 \cdot k_2 (v_2 - v_1) \cdot X \cdot T$$

while the total number per time and distance unit is as above.

Example 35. Within the time-distance intervals $X = 1$ km and $T = 1$ min there are two groups of vehicles.
    For the first group:

$$v_1 = 90 \text{ km/h}; \quad q_1 = 600 \text{ veh/h}; \quad k_1 = \frac{600}{90} = 6.67 \text{ veh/km},$$

while for the second group:

$$v_2 = 110 \text{ km/h}; \quad q_2 = 500 \text{ veh/h}; \quad k_2 = \frac{500}{110} = 4.55 \text{ veh/km}.$$

The total number of overtakings within this time-distance window is

$$M_p^a = k_1 \cdot k_2 \cdot X \cdot T \cdot (v_2 - v_1) = 6.67 \cdot 4.55 \cdot 1 \cdot \frac{1}{60} \cdot 20 = 10.$$

II.3.2 Partly Constrained Traffic

Traffic in which not all desired overtakings are possible is called partly constrained traffic (see Sect. II.2.6.3.1). The mathematical description is difficult and not yet completed. What follows is therefore intended to introduce and to illustrate the relevant concepts.

II.3.2.1 Queueing Theory Models for Two-Lane Rural Roads

To describe traffic flow behaviour in that part of the fundamental diagram where partly constrained flow occurs, queueing theory models have been developed. The service counters of the queueing system represent the vehicles to be overtaken. The arrivals at the service counter consist of the faster moving vehicles, which queue up behind the vehicle considered to await an opportunity to overtake. The service time at the counter corresponds to the time that elapses before an acceptable time gap for overtaking arises.
    Mathematicl derivations of queueing theory models are not included in this book. Instead, the relevant literature will be referred to. The essence of these models will be described using two queueing theory models.
    Consider traffic flow on a two-lane rural road. Both traffic streams (one in each direction) are assumed to be in stationary equilibrium. It is further assumed that the headways in both streams are exponentially distributed. The ability for any particular vehicle in one of the streams to be overtaken is limited, namely there must be a sufficiently large gap (measured in time units) in the opposing flow to

permit overtaking. This vehicle corresponds to a service point whose operation is determined by the opposing flow. It can be shown that the arrival rate at the service point has a Poisson distribution. Queueing discipline ensures "first in - first out" (FIFO). The queueing system is therefore of the type M/G/1 (M: Poisson arrivals, G: arbitrarily distributed service times, 1: one service point). The form of the distribution of the service times depends on the opposing traffic. As a consequence of the assumption of stationary equilibrium, both the distribution of the arrival and service times are independent of time.

All gaps in the opposing traffic greater than headway t_g are sufficient for overtaking. If more than one vehicle wishes to overtake in the same gap, a certain time must be allowed for the subsequent vehicles to move up in the queue before overtaking. This particular representation of the overtaking of a single vehicle on a two-lane rural road clearly resembles certain models used for the calculation of the capacity of priority intersections. The stream of vehicles wishing to overtake corresponds to a subsidiary flow, while the opposing traffic corresponds to the flow on a major road.

II.3.2.2 The Kinetic Theory

As in Sect. II.2.1, the quantity $\varkappa(x,t)\,dx$ can be interpreted under certain assumptions[1] as the probability that a vehicle is located at time t in the arbitrarily small interval $(x, x+\Delta x)$. Let $G(v_w)$ be the distribution function of the desired speed. Then the probability that a vehicle with desired speed (v_w, v_w+dv) is located at time t in the interval $(x, x+dx)$ is given by

$$f_w(x,t,v_w)\,dx\,dv = \varkappa(x,t)\,dx\,dG(v_w). \tag{II.150}$$

By definition the actual speed distribution $G(v)$ deviates from the desired speed distribution $G(v_w)$ in partly constrained traffic. Thus, in analogy with Eq. (II.150),

$$f(x,t,v)\,dx\,dv = \varkappa(x,t)\,dx\,dG(v). \tag{II.151}$$

Using the definition of concentration given in Sect. II.2.1

$$\int_{v_w=0}^{\infty} f_w(x,t,v_w)\,dv = \int_{v=0}^{\infty} f(x,t,v) = \varkappa(x,t) \int_{v=0}^{\infty} dG(v) = \varkappa(x,t). \tag{II.152}$$

Let us consider the question of how f changes in response to changes of the concentration and/or speed with time. Changes in f, in terms of $\delta f/\delta t$, can be thought of consisting of three parts:

1. Let us consider vehicles with speeds in the interval $(v, v+dv)$. As is shown by Fig. II.62, all vehicles which are located at time t in the interval $(x-dx, x)$ have by time $t+dt$ passed the location x. This number of vehicles is $f(x,t,v)\,dx\,dv$. Therefore, setting $dx=vdt$, the number of vehicles entering the differential element in Fig. II.62 is

$$f(x,t,v)\,dx\,dv = v\,f(x,t,v)\,dt\,dv.$$

---

1 Whether these assumptions are valid is here not discussed.

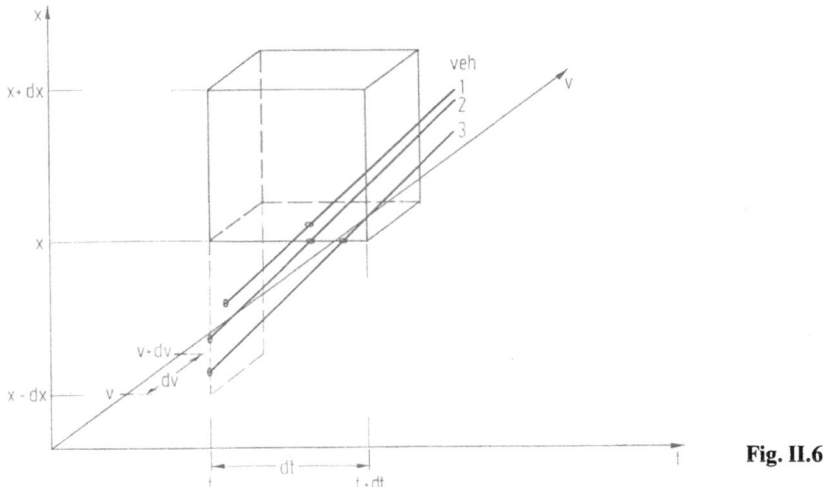

Fig. II.62

Correspondingly, the number of vehicles which leave the differential element during time dt at location $x + dx$ is

$$v\,f(x+dx,t,v)\,dt\,dv.$$

The change in the number of vehicles in differential element dx, dt, dv over the time interval Δt is just the difference between the vehicles entering and leaving. Thus

$$\frac{\delta f}{\delta t}\,dt\,dx\,dv = v\,f(x,t,v)\,dt\,dv - v\,f(x+dx,t,v)\,dt\,dv$$

from which

$$\frac{\delta f}{\delta t} = -v\,\frac{f(x+dx,t,v)-f(x,t,v)}{dx}\,\frac{\delta f}{\delta t} = -v\,\frac{\delta f(x,t,v)}{\delta x}. \tag{II.153}$$

Since $v = $ const, there is a change in concentration. As will be thoroughly explained in the discussion of continuum theory, Sect. II.3.3.3.1, such a change of the concentration over time is only possible if, at the same time, the intensity changes over distance. Let $\delta f/\delta t_{\text{cont}}$ (cont = continuum theory) denote that part of $\delta f/\delta t$ which is given by Eq. (II.153).

2. The interaction effect is the label applied, when vehicles are prevented by the presence of other, slower vehicles from maintaining their desired speeds. The expected number of vehicles which will interact with a vehicle having speed v_0 is equal to the product of the expected number of vehicles having $v > v_0$ with the probability $[1-P]$ that overtaking is not possible. The overtaking probability P is itself a function of the quantity of traffic, the directional split on two-way roads, the road characteristics (e.g. insufficient sight distance) etc. v. For the purpose of the discussion let us simply assume it to be constant. If $\varkappa(x,t)$ changes sufficiently slowly, then the number of vehicles per unit time having $v > v_0$ is approximately

$$E[q_p^p(v_0)] = \int_{v_0}^{\infty} (v-v_0)\varkappa(x,t)\,dG_m(v)$$

[see Eq.(II.125)]. Setting $f(x,t,v)dv = \varkappa(x,t)dG(v)$ [see Eq.(II.151)] we obtain

$$E[q_p^p(v_0)] = \int_{v_0}^{\infty} (v-v_0)f(x,t,v)dv.$$

Overall, then, there are

$$[1-P]\left[\int_{v_0}^{\infty} (v-v_0)f(x,t,v)dv\right]dt$$

vehicles interacting with a vehicle travelling at v_0 during a time dt. Because a total of $f(x,t,v_0)dxdv$ vehicles with a speed in the interval (v_0, v_0+dv) are located in dx at time t there is therefore a total of

$$[1-P]f(x,t,v_0)dx\,dv\,dt \int_{v_0}^{\infty} (v-v_0)f(x,t,v)dv \qquad (II.154)$$

interactions resulting from non-occurring overtakings of those vehicles with speed v_0. The portion of those vehicles having speed v_0, in the element (dt,dx,dv) increases per unit time by that number of vehicles which had initial speeds $v < v_0$ but were not able to overtake during the time dt in dx:

$$\frac{\delta f^+}{\delta t_{ww}}dx\,dt\,dv = [1-P]f(x,t,v_0)dt\,dx\,dv \int_{v_0}^{\infty} (v-v_0)f(x,t,v)dv. \qquad (II.155)$$

Vehicles having speed v_0 do more than just impede passive overtakings; they themselves are impeded from active overtaking by other vehicles having speed $v < v_0$. The probability is again assumed to be $1-P$. Since they are then reduced to $v < v_0$, they drop out of that portion of vehicles having speed v_0. Therefore, by analogy with Eq.(II.155),

$$\frac{\delta f^-}{\delta t_{ww}}dx\,dt\,dv = [1-P]f(x,t,v_0)dt\,dx\,dv \int_{0}^{v_0} (v_0-v)f(x,t,v)dv.$$

The total rate of change of the probability density function $f(x,t,v_0)$ due to interactions is then

$$\frac{\delta f}{\delta t_{ww}} = \frac{\delta f^+}{\delta t_{ww}} - \frac{\delta f^-}{\delta t_{ww}}$$

$$= [1-P]f(x,t,v_0)\left[\int_{v_0}^{\infty} (v-v_0)f(x,t,v)dv - \int_{0}^{v_0} (v_0-v)f(x,t,v)dv\right]$$

$$= [1-P]f(x,t,v_0) \int_{0}^{\infty} (v-v_0)f(x,t,v)dv. \qquad (II.156)$$

From Eq.(II.151) we can write the expression

$$\int_{0}^{\infty} vf(x,t,v)dv$$

as

$$\int_{0}^{\infty} v\varkappa(x,t)dG(v)$$

and, as shown in Sect. II.2.5.1, the result of this integration is

$$\varkappa(x,t) \int_{0}^{\infty} v\,dG(v) = \varkappa(x,t)E[V(x,t)].$$

Moreover, Eq. (II.152) gives the expression

$$\int_0^\infty f(x,t,v)\,dv = \varkappa(x,t)$$

with which Eq. (II.156) can be transformed to give

$$\frac{\delta f}{\delta t_{ww}} = [1-P]f(x,t,v_0)\varkappa(x,t)\,[E[V(x,t)]-v_0] \tag{II.157}$$

for the total temporal change of $f(x,t,v_0)$ due to interactions.

This equation assumes that speeds can instantaneously change from $v > v_0$ to v_0 and from $v_0 > v$ to v; further neglected in the derivation is the fact that impeded overtakings result in travelling queues of varying length, which themselves will of course influence the times and locations of interactions. We are making the important assumption that a vehicle is a point and that its actual physical length does not concern us here. More realistic results might be obtained by treating the interaction process as a queueing problem, in which the "servers" are moving, service times would be times during which vehicles are forced to travel with $v < v_0$.

3. Vehicles which are prevented from overtaking and therefore slow down will try to resume their desired speeds as soon as possible. Let the average time elapsing before a vehicle is able to resume its desired speed be T. In the literature, T is usually referred to as the relaxation time. It is assumed for simplicity that the temporal rate of change

$$\frac{\delta f}{\delta t_R} = -\frac{f(x,t,v)-f_w(x,t,v_w)}{T} \tag{II.158}$$

results (R = relaxation process). This is equivalent to assuming an exponential response from f to f_w. The negative sign in Eq. (II.158) results from the fact that the accelerating vehicles which leave a queue no longer have speed v_0. The vehicle at the front of the queue is able to travel at its desired speed while the following vehicles are forced to travel at $v_0 < v_w$.

The assumptions made here are crude. Relaxation time is in reality a random variable whose distribution function depends on the temperament of the drivers, on the time that an individual driver spends in the queue, on the acceleration capabilities of the vehicles, etc. The results derived here should serve to illustrate the problems of partly constrained traffic, but further work requires data which are not yet available.

Collecting together Eqs. (II.153), (II.157), and (II.158), the rate of change of $f(x,t,v)$ is therefore

$$\frac{\delta f}{\delta t} = \frac{\delta f}{\delta t_{cont}} + \frac{\delta f}{\delta t_{ww}} + \frac{\delta f}{\delta t_R}$$

$$= -v\frac{\delta f(x,t,v)}{\delta x} + [1-P]f(x,t,v_0)\varkappa(x,t)\,[E[V(x,t)]-v_0]$$

$$-\frac{f(x,t,v_0)-f_w(x,t,v_0)}{T}. \tag{II.159}$$

It follows that

$$\frac{\delta f(x,t,v)}{\delta t} + v\frac{\delta f(x,t,v)}{\delta x} = [1-P]f(x,t,v_0)\varkappa(x,t)\,[E[V(x,t)]-v_0]$$

$$- \frac{f(x,t,v) - f_w(x,t,v_w)}{T} \tag{II.160}$$

This approach resembles that of the Boltzmann equation in the kinematic theory of gases.

If it is not possible for queues to dissipate (i.e. for vehicles to leave queues by means of overtaking), then all vehicles travel in platoons. When overtaking becomes impossible, partly constrained traffic passes over into the constrained traffic regime.

II.3.2.3 A Multi-phase Model of Traffic Flow

A kinetic theory of traffic flow may be extended by combining a multi-phase model with the kinetic model. For example, the multi-phase model divides the traffic flow on a two-lane road into four phases. Each phase corresponds to a possible traffic flow state.

- Phase 1F: free flow in the right lane
- Phase 2F: free flow in the left lane
- Phase 1K: constrained flow in the right lane
- Phase 2K: constrained flow in the left lane.

The right hand side of Eq. (II.160) can be divided into a gain term G and a loss term $v \cdot f$. Thus, from Eq. (II.160) we obtain

$$\frac{\delta f(x,t,v)}{\delta t} + v\frac{\delta f(x,t,v)}{\delta x} = G(x,t,v) - v(x,t,v)\cdot f(x,t,v). \tag{II.161}$$

In the following, the density function $f(x,t,v,v_w)$ is considered. The possibility of finding a vehicle in distance interval $(x, x+dx)$ during time interval $(t+dt)$ that has an instantaneous speed of v and a desired free flow speed in the interval (v_w, v_w+dv_w) is denoted by $f(x,t,v,v_w)\,dx\,dt\,dv\,dv_w$.

This question can be specified for each of the four phases:

$$\frac{\delta f_{1F}(x,t,v_w)}{\delta t} + v_w\frac{\delta f_{1F}(x,t,v_w)}{\delta x} = G_{1F}(x,t,v_w) - v_{1F}(x,t,v_w)\cdot f_{1F}(x,t,v_w)$$

$$\frac{\delta f_{2F}(x,t,v_w)}{\delta t} + v_w\frac{\delta f_{2F}(x,t,v_w)}{\delta x} = G_{2F}(x,t,v_w) - v_{2F}(x,t,v_w)\cdot f_{2F}(x,t,v_w)$$

$$\frac{\delta f_{1K}(x,t,v,v_w)}{\delta t} + \frac{\delta f_{1K}(x,t,v,v_w)}{\delta x}$$
$$= G_{1K}(x,t,v,v_w) - v_{1K}(x,t,v,v_w)\cdot f_{1K}(x,t,v,v_w)$$

$$\frac{\delta f_{2K}(x,t,v,v_w)}{\delta t} + \frac{\delta f_{2K}(x,t,v,v_w)}{\delta x}$$
$$= G_{2K}(x,t,v,v_w) - v_{2K}(x,t,v,v_w)\cdot f_{2K}(x,t,v,v_w).$$

The transitions between the different phases are defined by transition rates. In the following, it will be assumed that the vehicles achieve the transition from one phase to another in infinitesimally small steps. (This supposes unlimited acceleration and braking capabilities.)

Figure II.63 shows the possible transitions between the phases:

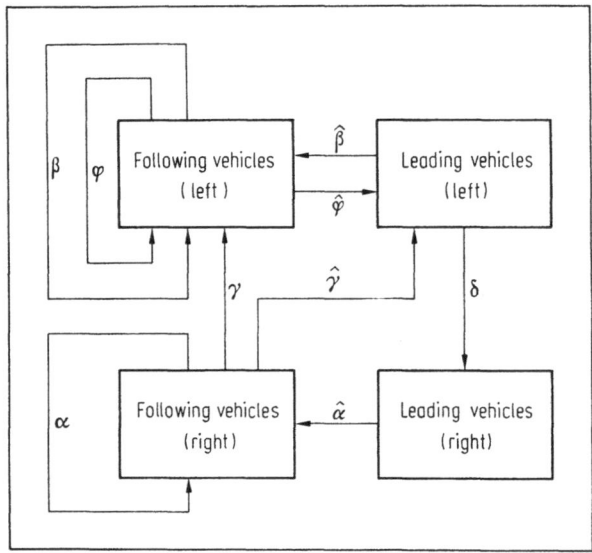

Fig. II.63. (From [66])

The individual transition phases are in detail:

$\hat{\alpha}$: A vehicle travelling with desired free flow speed v_w in the right lane approaches another vehicle travelling in the same lane at $v < v_w$, and thereafter travels at v maintaining a safety distance of $a = a(v)$.

$\hat{\beta}$: As for $\hat{\alpha}$, but for the left lane.

α: A vehicle in the right lane travelling at $v < v_w$ notices that the vehicle in front is driving at $v_1 < v$, approaches it, slows to v_1 and then follows it at a distance of $a(v_1)$.

β: As for α, but for the left lane.

γ: A vehicle in the right lane travelling at $v < v_w$ finds a sufficiently large gap behind another vehicle travelling at $v < v_1$, in the left lane, and changes lane. There it maintains a speed of v_1.

$\hat{\gamma}$: A vehicle travelling in the right lane with speed $v < v_w$ finds a gap in the left lane that is greater than the safety distance corresponding to its desired free flow speed. It changes lane and continues at its desired free flow speed.

δ: A vehicle travelling freely in the left lane discovers a gap in the right lane within which it drives for a certain period at its desired free flow speed.

ψ: As a consequence of the leading vehicle of a platoon moving over to the left lane, the vehicle now in front of it accelerates to $v_1 < v_w$. It follows at v_1, maintaining a distance of $a(v_1)$.

$\hat{\psi}$: As a consequence of the leading vehicle of a platoon moving over to the left lane, the vehicle now in front of it accelerates to $v_1 \geq v_w$. It follows at its desired free flow speed v_w.

Thus the following system of differential equations is obtained:

$$\frac{\delta f_{1F}(x,t,v_w)}{\delta t} + v_w \frac{\delta f_{1F}(x,t,v_w)}{\delta x}$$

$$= G_{1F}(x,t,v_w) - v_{1F}(x,t,v_w) \cdot f_{1F}(x,t,v_w)$$

$$= f_{2F}(x,t,v_w) \cdot \delta(v_w) - f_{1F}(x,t,v_w) \int_0^{v_w} \alpha(v_w,v) dv$$

$$\frac{\delta f_{1K}(x,t,v_w)}{\delta t} + v \frac{\delta f_{1K}(x,t,v,v_w)}{\delta x}$$

$$= G_{1K}(x,t,v,v_w) - v_{1K}(x,t,v,v_w) f_{1K}(x,t,v,v_w)$$

$$= \int_0^{v_w} \alpha(\dot{v},v_1) f_{1K}(x,t,v_1,v_w) dv_1 + \alpha(v_w,v) f_{1F}(x,t,v_w)$$

$$- f_{1K}(x,t,v,v_w) \left\{ \int_0^{v} \alpha(v_1,v) dv + \int_v \gamma(v,v_1) dv_1 + \gamma(v,v_w) \right\}$$

$$\frac{\delta f_{2F}(x,t,v_w)}{\delta t} + v_w \frac{\delta f_{2F}(x,t,v_w)}{\delta x}$$

$$= G_{2F}(x,t,v_w) - v_{2F}(x,t,v_w) \cdot f_{2F}(x,t,v_w)$$

$$= \int_0^{v_w} f_{1K}(x,t,v,v_w) \cdot \hat{\gamma}(v,v_w) dv$$

$$+ \int_0^{v_w} f_{2K}(x,t,v,v_w) \hat{\psi}(v,v_w) dv$$

$$- f_{2F}(x,t,v_w) \cdot \left\{ \int \hat{\beta}(v_w,v) + \delta(v_w) \right\}$$

$$\frac{\delta f_{2F}(x,t,v,v_w)}{\delta t} + v \frac{\delta f_{2K}(x,t,v,v_w)}{\delta x}$$

$$= G_{2K}(x,t,v,v_w) - v_{2K}(x,t,v,v_w) f_{2K}(x,t,v,v_w)$$

$$= f_{2F}(x,t,v_w) \cdot \hat{\beta}(v_w,v)$$

$$+ \int_v^{v_w} f_{2K}(x,t,v_1,v_w) \beta(v_1,v) dv_1$$

$$+ \int_0^{v} f_{2K}(x,t,v_1,v_w) \Psi(v_1,v) dv_1$$

$$+ \int_0^{v} f_{1K}(x,t,v_1,v_w) \gamma(v_1,v) dv_1$$

$$- f_{2K}(x,t,v,v_w) \cdot \left\{ \hat{\psi}(v,v_w) + \int_0^{v} \beta(v,v_1) dv_1 + \int_v^{v_w} \psi(v,v_1) dv_1 \right\}$$

The calculation of the transition rates is complex, and will not be discussed further here.

A solution of the system of equations specified above is not feasible analytically. If only discrete steps in the value of speed are allowed, the system of equations may be solved numerically.

II.3.2.4 Observed Lane Changing Behaviour

A clear distinction should be drawn between overtaking as so far considered and lane changing. Overtaking has concerned only the passing of a slower vehicle by a faster vehicle. Whether overtaking utilised the neighbouring left or right lanes, or whether lane changing was involved, was not considered. These aspects depend on

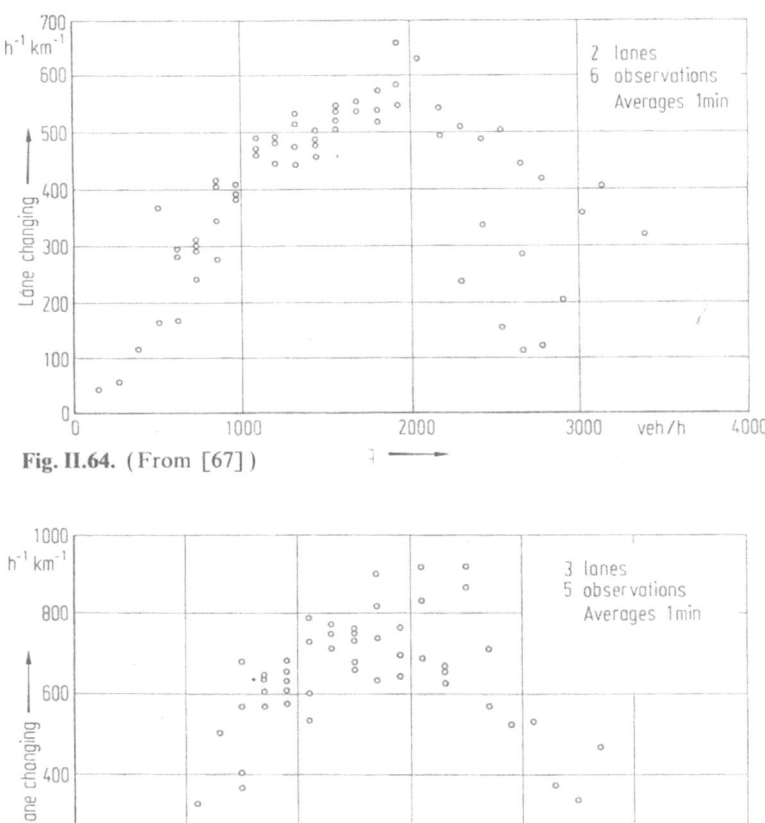

Fig. II.64. (From [67])

Fig. II.65. (From [68])

the traffic situation and the rules of behaviour. Even in the case of two-lane single carriageways, where the relationship between passing and lane changing is closest, a lane change may be associated with the overtaking of one or more vehicles.

To date it has not proved possible to obtain analytically the number of lane changes as a function of the traffic volume. This relationship can only be observed or simulated.

Observations made on a section of a German autobahn have clearly demonstrated a relationship between the rate of lane changing (per time and distance interval) and traffic volume. For one carriageway of an autobahn, the maximum rate of lane changing was observed to occur with a flow of 2 000 veh/h on a two-lane section (see Fig. II.64) and with flow of 3 300 veh/h on a three-lane section (see Fig. II.65). However, before carrying these results over to other countries, the differing traffic rules have to be considered.

Theoretical approaches to the description of the frequency of lane changing in general require such empirical relationships as input. These aspects are gone into later.

II.3.3 The Constrained Traffic

This section is concerned with those vehicles which, in a partly constrained traffic flow, or in the limiting case of constrained traffic, travel in platoons.

A sequence of vehicles is defined as two or more vehicles on a lane or a fixed track one behind the other. A platoon is one portion of a sequence of vehicles in which each vehicle except the first is forced to adjust its speed to that of the vehicle ahead of it. Thus a platoon can have as few as two vehicles so long as the second vehicle is forced to follow the first vehicle.

Platoons can be viewed as moving queues (see Sect. II.3.2). One, however, differentiates between the platoon leader and those vehicles which follow it.

II.3.3.1 Deterministic Spacing Models

Deterministic spacing models try to describe the spacing characteristics using kinematic variables (speed, deceleration, etc.). The advantage is that they are easy to deal with; their disadvantage lies in the necessary simplification of the model structure, particularly with the respect to the relationship between the spacing, and the driver observation-reaction process.

II.3.3.1.1 Constant Spacing

It will be assumed that

a) all vehicles travel with the same speed v.
b) all vehicles have the same length ($l_f = $ const), and
c) all vehicles maintain the same constant spacing $a = \Delta x - l_f = $ const (see Fig. II.66).

The model assumptions can be illustrated by imagining a conveyor belt, which can move at different speeds, on which, for example, bricks are placed at equal

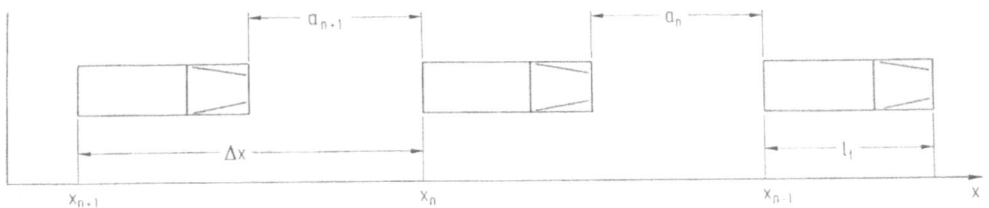

Fig. II.66

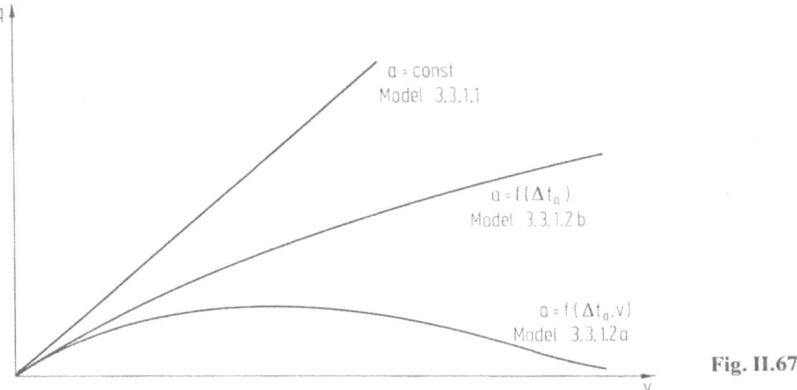

Fig. II.67

spacing. Then

$$k = 1/\Delta x \tag{II.162}$$

and

$$q = vk = v/\Delta x \tag{II.163}$$

(see Sect. II.2.5.1). The traffic volume in this case increases linearly with speed[1] (see Fig. II.67). Such a model is obviously unrealistic, not only because speeds are assumed to be all equal, but especially because the spacings are independent of the speed.

II.3.3.1.2 Spacing as a Function of Speed

An improved model (the total safety distance model) proceeds from the safety requirement that the distance a, or Δx, be sufficiently large to permit a vehicle to brake to a stop without causing a rear-end collision if the preceding vehicle, for any reason whatever, comes to a stop instantaneously. The required distance x referred to as the total safety distance consists (Fig. II.68) of

a) the vehicle length l_f,
b) the distance covered during the perception, decision and reaction time l_a,
c) the minimum possible braking distance l_b,
d) the reserve safety distance (at rest) l_s.

---

[1] This model is sometimes used as the basis of the argument, that speed limits would clearly reduce road capacities.

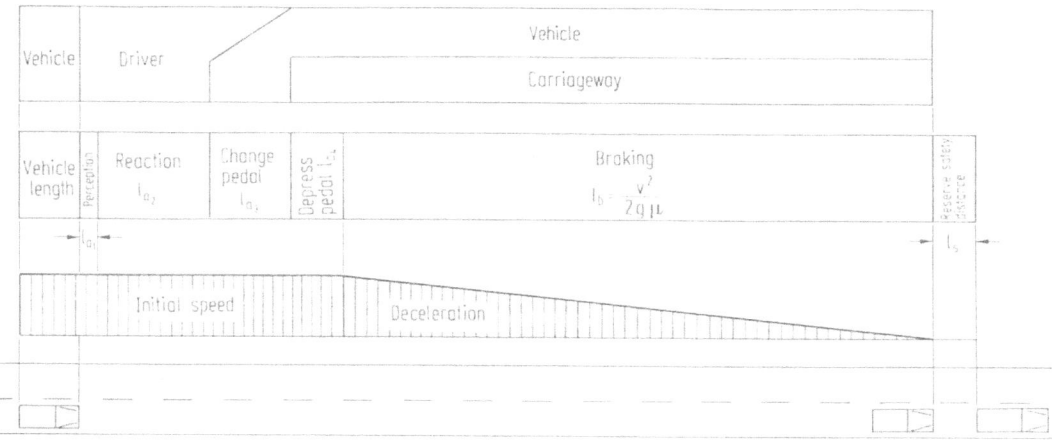

Fig. II.68. (From [79])

The overall reaction time, Δt_a, is the time from the appearance of an obstacle to the beginning of the deceleration and consists of

a) the perception time; this is the time that the driver needs to recognize that there is an obstacle,
b) decision time; this is the time that the driver needs to make a decision to stop immediately,
c) the time needed by the driver to move his foot to the brake pedal, and
d) the pedal time needed to push the brake pedal.

The last two times depend both on the driver and on the vehicle design. Since during the overall reaction time the speed remains nearly constant, $l_a = \Delta t_a \cdot v$. The braking distance l_b (from the onset of deceleration until the vehicle comes to a complete stop) depends on the speed, on the road, and on the vehicle. It can be calculated as

$$l_b = \frac{Gv^2}{2G_b g\mu}$$

for a level road, where μ is taken as being constant.

G = the weight of the vehicle,

G_b = the weight supported by those axles which are equipped with brakes,

μ = the coefficient of friction between the tires and the road surface, in reality a function of v, as shown in Fig. II.69,

g = acceleration due to gravity.

In general, $G = G_b$ and thus $l_b = v^2/2g\mu$. Therefore, $\Delta x = l_f + l_a + l_b + l_s$.

In addition to the total safety distance model [denoted in what follows as Model (a)], the following model is also possible. Assume that two vehicles travel together at approximately the same speed, as is frequently the case in heavy traffic.

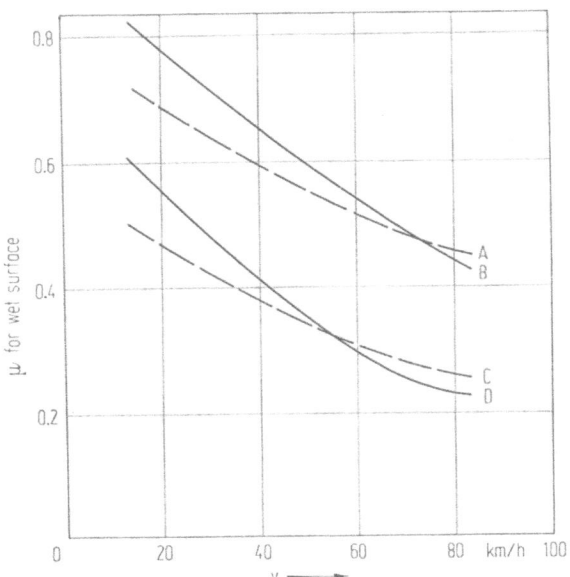

Fig. II.69. A, good surface with rough open texture; B, good surface with smooth texture; C, inadequate surface with rough open texture; D, inadequate surface with smooth texture. (From [83])

Their braking distances are nearly the same. If the first vehicle brakes, then the second vehicle needs only the distance it covers during the overall reaction time with unreduced speed. This is the reaction time distance model, denoted in the following as Model (b).

Model (a): Considering the entire braking distance, we have

$$\Delta x = l_f + l_a + l_b + l_s = l_f + \Delta t_a v + \frac{v^2}{2g\mu} + l_s.$$

When the total braking distance is included, combining the two quantitites l_f and l_s into one, we obtain:

$$\Delta x = \Delta t_a v + \frac{v^2}{2g\mu} + l_{sf}.$$

If one makes a graph of the traffic volume as a function of speed, as computed using Model (a), then one obtains a curve having a maximum (Fig. II.67). It follows, that there is therefore a speed at which a maximum traffic volume will occur; this speed is called v_{opt}.

Substituting the quantity

$$\Delta x = l_{sf} + \Delta t_a v + \frac{v^2}{2\mu g}$$

for the distance-headway in Eq. (II.163), thereby assuming that all vehicles have the same speed-dependent spacing, we obtain

$$q = \frac{v}{l_{sf} + \Delta t_a v + \dfrac{v^2}{2\mu g}} \, .$$

The corresponding average time-headway is

$$\bar{z} = \frac{1}{q} = \frac{l_{sf}}{v} + \Delta t_a + \frac{v}{2\mu g} \, .$$

Since $\bar{z}_{min}$ corresponds to q_{max}, let us find v_{opt} by deriving an expression for $\bar{z}_{min}$:

$$\frac{d\bar{z}}{dv} = l_{sf} \left(-\frac{1}{v^2} \right) + \frac{1}{2\mu g} \, .$$

Setting $d\bar{z}/dv = 0$ yields the desired results

$$v_{opt} = \sqrt{2\mu g l_{sf}}; \quad \bar{z}_{min} = \Delta t_a + 2\sqrt{\frac{l_{sf}}{2\mu g}} \, .$$

The optimal speed is therefore not dependent on Δt_a but is only a function of the coefficient of friction and l_{sf}.

Model (b): Since the distance-headway in this model does not include the braking distance, but only the distance covered during the overall reaction time,

$$\Delta x = \Delta t_a v + l_{sf} \, .$$

As Fig. II.67 shows, the resulting function $q = q(v)$ has no maximum for finite v but rather as an asymptote at a value $q = 1/\Delta t_a$.

Model (a) can be generalized by writing the coefficient of friction as a function of speed (see Fig. II.69) so that

$$\Delta x = \Delta t_a v + \frac{v^2}{2g\mu(v)} + l_{sf} \, .$$

This model can be extended so as to include the possibility that the coefficients of friction for the two vehicles are different, thereby resulting in unequal braking distances. In this case,

$$\Delta x = \Delta t_a v + \frac{v^2}{2g} \left(\frac{1}{\mu_1} - \frac{1}{\mu_2} \right) + l_{sf} \, .$$

Furthermore, observed distance-headways can be described by expressions of the form

$$\Delta x = cv^n + l_{sf} \, ,$$

where cv^n are empirically determined constants. Since $q = k \cdot v$ and $k(v) = 1/a(v)$, we have $q(v) = v/a(v)$, demonstrating that the form of function $q(v)$ is determined by $a(v)$. The form of the fundamental diagram $q(k)$ is therefore also determined by $a(v)$.

Example 36. One method by which the distance [m] to the vehicle in front may be set so that the headway remains above the legal minimum is to divide the speed [km/h] by two. Hence $a(v)\,[m] = 0.5 \cdot v\,[km/h]$ and

$$q(v) = \frac{v}{0.5v} = 2 \left[\frac{km \cdot veh}{h \cdot m} \right] = 2000 \text{ veh/h.}$$

Thus, q is independent of the value of v (see Fig. II. 70).
Therefore, q is also independent of k.

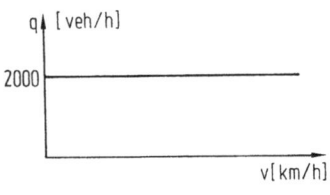

Fig. II.70

Example 37. In a lane, all vehicles travel at the same speed and with the same total safety distance. Assuming a vehicle length of 4.5 m, a reserve safety distance of 0.5 m, a maximum braking deceleration of 6 m/s^2, and omitting any reaction times and other lags, we obtain

$$\Delta x = 4.5 \text{ m} + 0.5 \text{ m} + \frac{v^2}{2 \cdot b} \,.$$

Then for $v = 80$ km/h

$$\Delta x_1 = 5 \text{ m} + \frac{80^2}{3.6^2 \cdot 2 \cdot 6} = 46.14 \text{ m.}$$

Thus,

$$k_1 = \frac{1000}{46.14} = 21.65 \text{ veh/km}$$

hence

$$q_1 = 21.65 \cdot 80 = 1732 \text{ veh/h.}$$

For $v = 100$ km/h

$$\Delta x_2 = 5 \text{ m} + \frac{100^2}{2.6^2 \cdot 2 \cdot 6} = 69.31 \text{ m.}$$

Therefore

$$k_2 = \frac{1000}{69.31} = 14.43 \text{ veh/km}$$

and

$$q_2 = 14.43 \cdot 100 = 1443 \text{ veh/h.}$$

This example illustrates that within a certain range of speeds, traffic flow increases rather than decreases with the imposition of a speed restriction (see also Fig. II.67).

Example 38. A firm offers a new, automatic means of transport that consists of small cabins mounted on a special track. The travelling speed is 10 m/s and the cabins are 2 m long.

The vehicles have a braking deceleration of $b_{max} = -5\,m/s^2$ and no reaction time on account of an automatic control system. The braking distance is

$$l_b = \frac{v^2}{2b} = \frac{10^2}{2\cdot 5} = 10\,m$$

hence the necessary distance is $a = 10 + 2 = 12\,m$. Therefore

$$q = \frac{v}{a} = \frac{10}{12}\cdot 3.6\cdot 1000 = 3000\;veh/h.$$

In order to raise the performance of the system, a positive residual speed on collision, v_R, is allowed (see Example 10). Thus

$$l_b = \frac{v^2 - v_R^2}{2b} = \frac{100 - v_R^2}{10} = 10 - \frac{v_R^2}{10}$$

and hence

$$a(v_R) = 12 - \frac{v_R^2}{10}.$$

Figure II.71 shows $q = v/a(v_R)$ as a function of v_R.

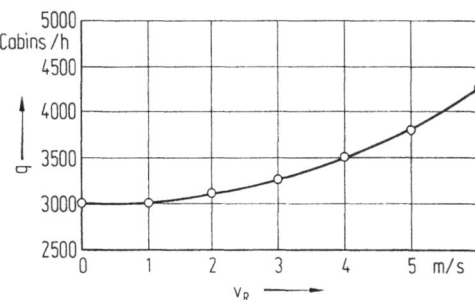

Fig. II.71

II.3.3.1.3 Car Following Models

Clearly, the assumption that, for any given speed there is a fixed headway, even though it is speed-dependent, is still unrealistic. But it is possible neither to estimate the necessary distance exactly, nor to maintain it exactly, even if, somehow, it were estimated. In reality, following drivers try to conform with the preceding vehicles' behaviour (see also Sect. II.3.3.2). This process is called car following and is based on a cycle of stimulus and response. The nature of the response is an acceleration or a deceleration, delayed by an overall reaction time T. This process therefore resembles a feedback control process in which oscillations may occur. Such oscillations can sometimes lead to various kinds of instabilities in the traffic flow, which in turn can lead to accidents.

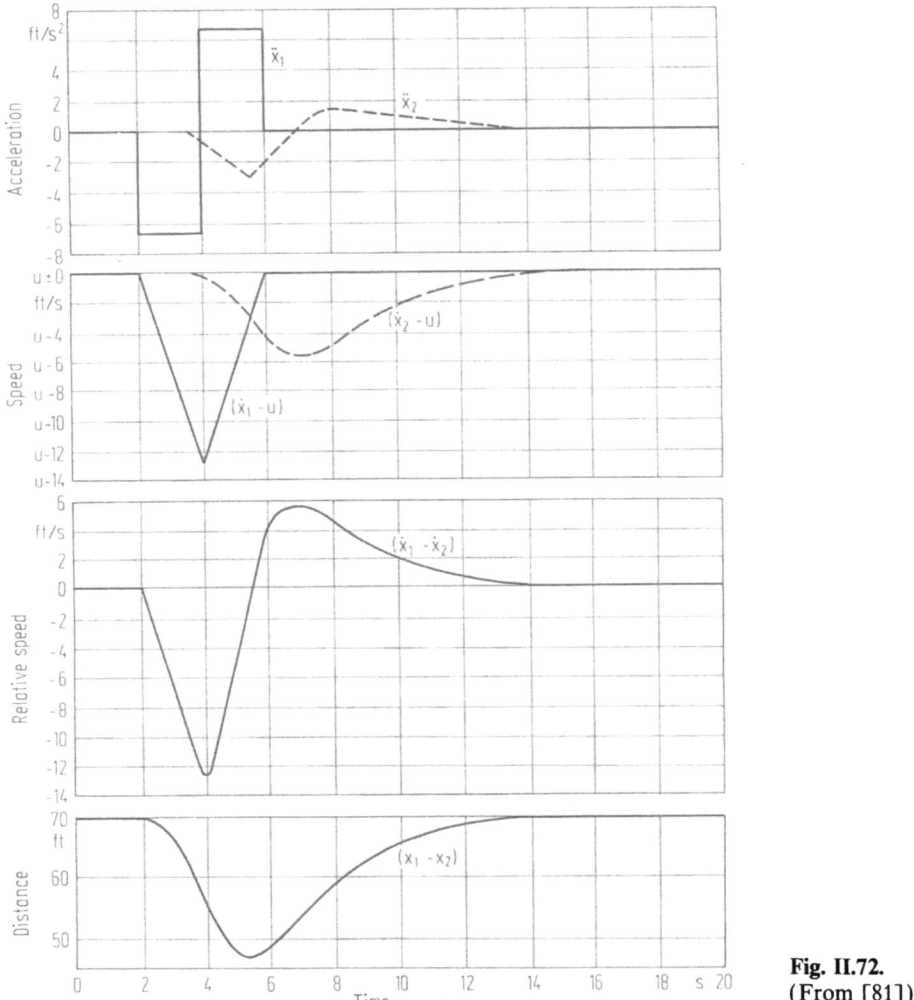

Fig. II.72.
(From [81])

There are two types of flow instability:

a) Local instability is defined as the situation in which a disturbance does not die out but rather increases with time. A disturbance is defined as a change in a distance-headway resulting from the change in speed of the leading vehicle (Figs. II.72 and II.73).

b) Asymptotic instability is defined as the situation in which a disturbance grows in magnitude as it propagates from vehicle to vehicle (Figs. II.74 and II.75).

Figure II.72 illustrates the phenomenon of car following for two vehicles in which the second vehicle follows the first according to Eq. (II.164). In this example the response is damped. In Fig. II.73, the first two cases ($C=0.50$ and $C=0.80$; the definition for C follows later) appear as damped oscillations, the third ($C=1.57$)

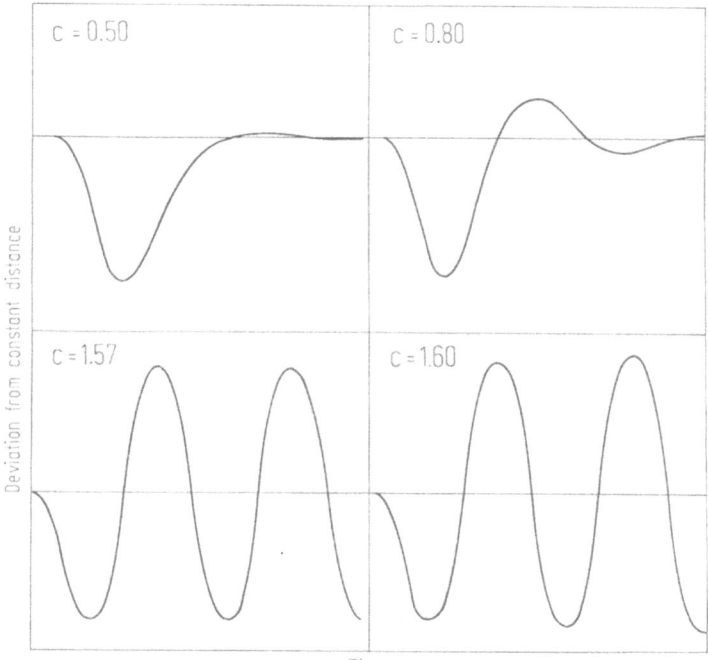

Fig. II.73. (From [81])

is an undamped oscillation having constant amplitude and the fourth ($C=1.60$) is a growing oscillation; it is locally unstable. Figure II.74 illustrates, for a platoon of eight interacting vehicles, first asymptotic stability in which the response is damped and non-oscillatory ($C=0.368$), second damped oscillation ($C=0.50$), and third asymptotic instability ($C=0.75$). Figure II.75 makes clear the consequences of asymptotic instability on the trajectories of a platoon of vehicles. Here these trajectories intersect either a collision or an emergency braking must result.[1]

Which changes in what variables the driver of the following vehicle responds to is difficult to determine. In the past an entire family of car following models has been investigated based on the following general form:

$$\ddot{x}_{n+1}(t+T) = \alpha[\dot{x}_n(t) - \dot{x}_{n+1}(t)]$$

where

$$\alpha = \frac{c\dot{x}_{n+1}^m(t+T)}{[x_n(t) - x_{n+1}(t)]^l} \tag{II.164}$$

yielding

$$\ddot{x}_{n+1}(t+T) = c\dot{x}_{n+1}^m(t+T)\frac{[\dot{x}_n(t) - \dot{x}_{n+1}(t)]}{[x_n(t) - x_{n+1}(t)]^l}. \tag{II.165}$$

---

1 Such emergency braking explains the phenomenon in traffic of "congestion from nowhere".

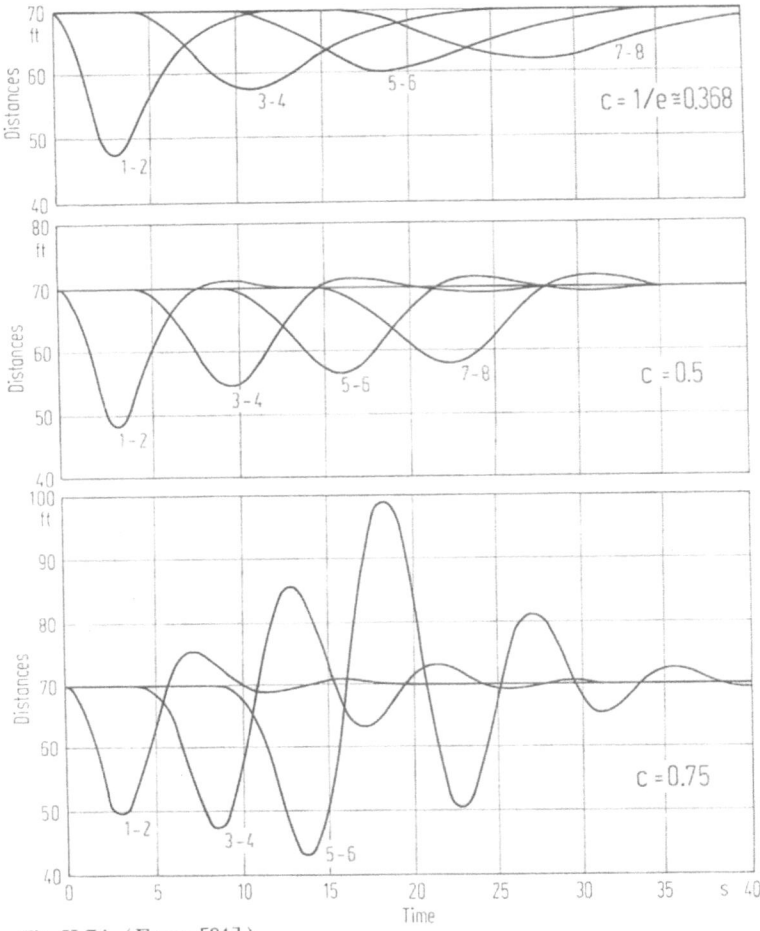

Fig. II.74. (From [81])

According to Eq. (II.165), the following vehicle changes its speed in proportion to the speed difference but delayed by time T. The proportionality factor α is called the sensitivity. It can be assumed to be

a) constant (including the special cases of α has different constant values for acceleration and deceleration) $(m=0, l=0)$
b) dependent on the distance-headway at time t $(m=0, l\neq0)$
c) dependent on the speed of the following vehicle at time $t+T$ $(m\neq0, l=0)$
d) dependent on both speed and distance-headway $(m\neq0, l\neq0)$.

Other approaches besides Eq. (II.165) have also been tried. They resemble Eq. (II.165) to the extent that they generally attempt to describe car following as a function of kinematic variables.

Stability computations for the non-linear models (b) through (d) are very difficult to carry through. The analysis of local stability for the linear model (a) produces the result (see Fig. II.73) that

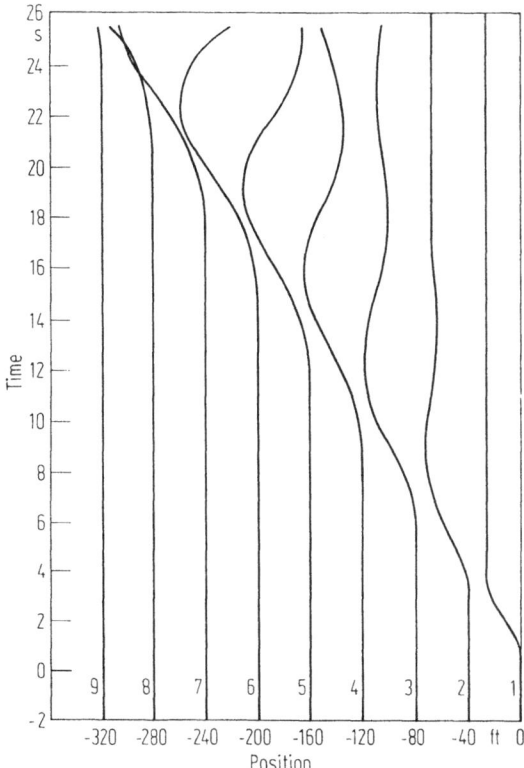

Fig. II.75. (From [81])

for $C=\alpha T\leqq 1/e$ (≈ 0.368)	the response is in general stable; the response to a pulse input (see Fig. II.72) is non-oscillatory and damped,
for $1/e<C<\pi/2$	the response is in general stable; the response to a pulse is oscillatory and damped,
for $C=\pi/2(\approx 1.57)$	the response is on the stability boundary; the response to a pulse is an oscillation with constant amplitude, and
for $C>\pi/2$	the response is unstable; the response to a pulse is an oscillation with growing amplitude.

Asymptotic stability requires that $C\leqq 0.5$.

If the motion of the first vehicle and all initial conditions of the platoon (distance-headways and speeds at $t=0$) are known, the trajectories of the following vehicles, for the model of Eq. (II.165), can then be determined:

a) through analytical calculations (at least for $\alpha=$ const), if the motion of the first vehicle is given as a continuous function,
b) through the iterative application of numerical methods (manually or by machine calculation),
c) graphically (not yet possible for all cases).

The graphical approach will be explained using the case $a=0$ as an example.

Example 39. Let us assume:

1. The speed-time profile of vehicle n is a sinewave (Fig. II.76)

$$x_n(t) = 10 \sin 0.4\, t\ \text{m/s} \quad \text{for } 0 \leq t \leq 7.84\ \text{s}.$$

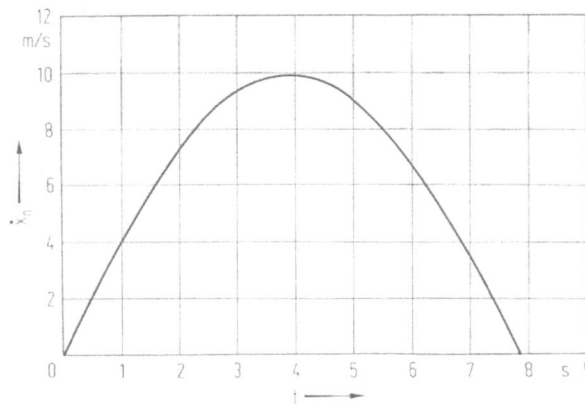

Fig. II.76

2. Let $T = 1$ s, $\alpha = 1$ s^{-1}. The acceleration (and with it the speed) can then be determined graphically as follows. Let us use a coordinate system in which the ordinate is the speed difference of the two vehicles and the abscissa is time (Fig. II.78). Plot a line parallel to the ordinate which passes through $t = 1/\alpha$. The slope of the straight line from the origin of a point of the line drawn through $t = 1/\alpha$ gives the acceleration of vehicle $n + 1$ at time t (where the relative speed is for the time $t - T$). The resulting slope

$$\alpha(\dot{x}_n - \dot{x}_{n+1})_{t-T}$$

responds to the right side of the car following Eq. (II.164) and thus yields the desired acceleration. Assuming that the acceleration remains constant over some time interval, the line segments in Fig. II.78 may be transferred to Fig. II.77 to yield the curve shown. The smaller the time interval the more accurate this construction will be. The resulting graphical solution (= speed-time profile of vehicle $n + 1$) is indicated in Fig. II.77 as a dashed line. For comparison, the exact solution as can be determined through analytical or numerical calculation, is shown as a solid line in the same figure.

II.3.3.1.4 Macroscopic Verification of Microscopic Models

II.3.3.1.4.1 Integration of the Car Following Equation

Models which, like car following models, attempt to explain traffic phenomena on the basis of the behaviour of its individual elements — single vehicles — are called *microscopic models*. In contrast, when traffic flow phenomena are described through parameters which characterize the collective traffic properties (e.g. traffic volume and traffic density), the resulting models are labelled *macroscopic models*. To the extent that both types of models describe the same phenomena, one should be derivable from the other.

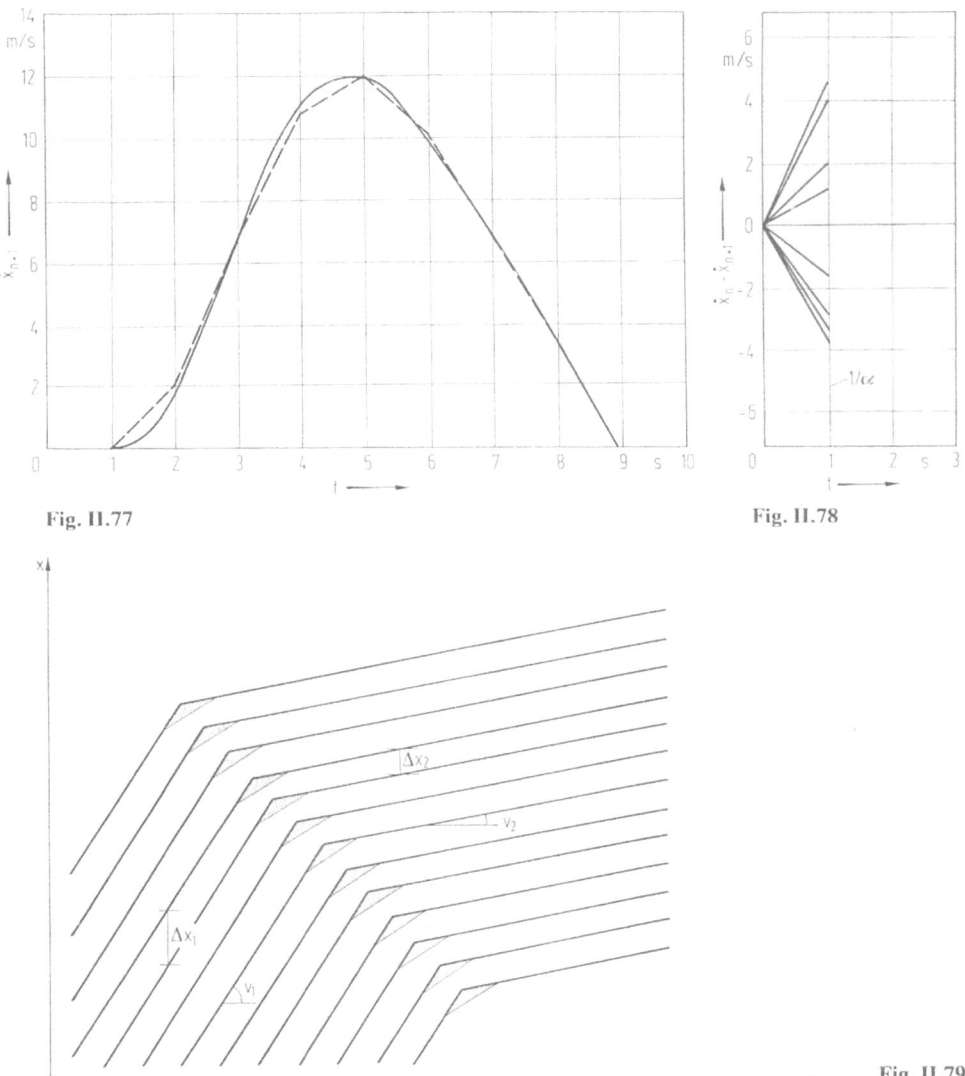

Fig. II.77

Fig. II.78

Fig. II.79

A platoon moves in a stationary fashion when, on the average (microscopically speaking), the speed $v = \dot{x}$ and the distance-headway Δx of a vehicle does not change with time. The resulting macroscopic relationship are

$$k = 1/\Delta x = \text{const} \quad \text{and} \quad q = vk = \text{const}.$$

For a stationary vehicle stream the equations of a single vehicle follow from Eq. (II.164)

$$\ddot{x}_{n+1}(t+T) = 0$$

and

$$\dot{x}_n(t) - \dot{x}_{n+1}(t) = 0.$$

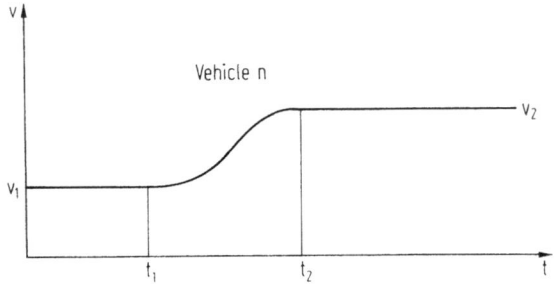

Fig. II.80

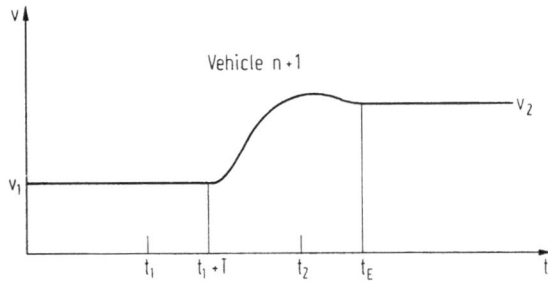

Fig. II.81

Now let the first vehicle change its speed from $v_1^{(n)}$ to $v_2^{(n)}$. The parameters α and T are assumed to be constants, such that the equations of motion are stable. In this case the response of the following vehicle is such that its speed after a certain time also becomes $v_2^{(n+1)} = \text{const} = v_2^{(n)}$ with a corresponding distance-headway $\Delta x_2 (= \Delta x_1) = \text{const}$. This transition can be described using Eq. (II.164).

Vehicle n changes its speed according to Fig. II. 80 from $v_1^{(n)} = \text{const}$ to $v_2^{(n)} = \text{const}$. Vehicle $n+1$ responds approximately as shown in Fig. II.81. $v_2^{(n+1)}$ is then calculated as

$$v_2^{(n+1)} = v_1^{(n+1)} + \int_{t_1+T}^{t_E} \ddot{x}_{n+1}(t)\,dt.$$

Since in the region where $v_1^{(n+1)} = v_1^{(n)} = v_1 = \text{const}$

$$\int_0^{t_1+T} \ddot{x}_{n+1}(t)\,dt = 0$$

and in the region where $v_2^{(n+1)} = v_2^{n} = v_2 = \text{const}$

$$\int_{t_E}^{\infty} \ddot{x}_{n+1}(t)\,dt = 0$$

we can change the limits of integration and obtain

$$v_2 = v_1 + \int_0^{\infty} \ddot{x}_{n+1}(t)\,dt.$$

Since the parameter T (Eq. II.164) disappears, we can in general write

$$v_2 = v_1 + \alpha \int_0^{\infty} [\dot{x}_n(t) - \dot{x}_{n+1}(t)]\,dt$$

$$v_2 - v_1 = \alpha\{x_n(\infty) - x_{n+1}(\infty) - [x_n(0) - x_{n+1}(0)]\}.$$

Setting

$$x_n(\infty) - x_{n+1}(\infty) = \Delta x_2 = 1/k_2$$

and

$$x_n(0) - x_{n+1}(0) = \Delta x_1 = 1/k_1$$

(see Fig. II.79), we obtain

$$v_2 - v_1 = \alpha\left(\frac{1}{k_2} - \frac{1}{k_1}\right).$$

We have now derived a macroscopic relationship between speed and density. Given the initial condition $v_1 = 0$ we obtain $k_1 = k_{max}$ (see Sect. II.2.5.3.3).

If this equation is assumed to describe the transition from the state $v_1 = 0$ (platoon at rest) to some other state $v_2 = v_1$ then

$$v = \alpha\left(\frac{1}{k} - \frac{1}{k_{max}}\right) \tag{II.166}$$

and

$$q = q(k) = vk = \alpha\left(1 - \frac{k}{k_{max}}\right). \tag{II.167}$$

But this is just an equation for the fundamental diagram. Since this is a linear relationship between q and k (Fig. II.82), it obviously does not fulfill the required boundary conditions listed in Sect. II.2.5.3.3:

for $k=0$, $q=\alpha(\neq 0)$:   boundary condition 1 is not fulfilled

for $k=k_{max}$, $q=0$:   boundary condition 2 is fulfilled

for $k=0$, $v=\infty(\neq v_w)$:   boundary condition 3 is not fulfilled

for $k=k_{max}$, $v=0$:   boundary condition 4 is fulfilled

$$\lim_{k \to 0} \frac{dv}{dk} = \frac{\alpha}{k^2} = \infty(\neq 0):$$ boundary condition 5 is not fulfilled

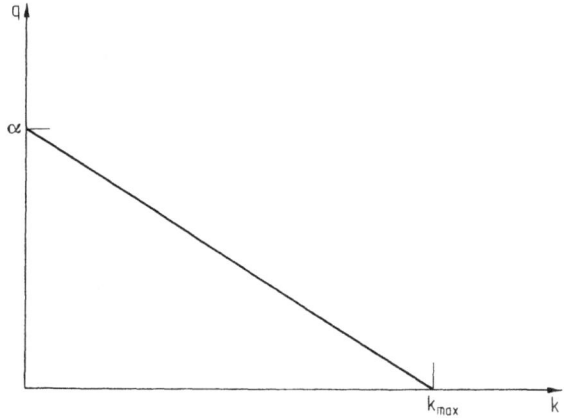

Fig. II.82

It should be noted that, because of the basic assumption of car following models (namely that drivers are always under the influence of leading vehicles), one should not have great hopes for a realistic macroscopic description of light traffic. Nevertheless, the great benefit of a model like Eq. (II.164) with $\alpha = \text{const}$ is that it permits the phenomena of platoon stability to be illustrated reasonably simply.

Since the reaction time T disappears through an integration (see page 138) Eq. (II.166) can be derived directly from Eq. (II.164):

$$\ddot{x}_{n+1}(t+T) = \alpha[\dot{x}_n(t) - \dot{x}_{n+1}(t)]$$

$$\int \ddot{x}_{n+1}(t+T)dt = \alpha \int [\dot{x}_n(t) - \dot{x}_{n+1}(t)]dt$$

$$\dot{x} = \alpha \Delta x + C$$

or

$$v = \frac{\alpha}{k} + C.$$

With the initial condition $k = k_{max}$ for $v = 0$, C is found to be

$$C = -\alpha/k_{max}$$

so that

$$v = \alpha \left(\frac{1}{k} - \frac{1}{k_{max}} \right)$$

as above [Eq. (II.166)].

II.3.3.1.4.2 The Influence of Exponential Terms in the Car Following Equation on the Shape of the Fundamental Diagram

The integration of the general car following model of Eq. (II.165) is substantially more complicated. This integration yields an equation of the form

$$f_m(v) = -\alpha_0 f_1(\Delta x) + \beta. \tag{II.168}$$

The quantities $f_m(v)$ and $f(\Delta x)$ are represented together by $f_p(z)$. Thus p stands for m or 1 and z for v or Δx. It can be shown that

1a) $f_p(z) = z^{1-p}$ for $p \neq 1$
1b) $f_p(z) = \ln z$ for $p = 1$.

Furthermore

2a) $\beta = f_m(v_w)$ for $m > 1, 1 \neq 1$, or $m = 1, 1 > 1$

    [v_w = desired speed (see Sect. II.2.5.3)]

2b) $\beta = \alpha_0 f_1(\Delta x_{min})$ for all combinations of m
                                             and l, except for $m = 1, 1 < 1$

    ($\Delta x_{min} = 1/k_{max}$).

Consider now a model with $m = 0, 1 = 1$:

$$\ddot{x}_{n+1}(t+T) = \alpha_0 \frac{\dot{x}_n(t) - \dot{x}_{n+1}(t)}{x_n(t) - x_{n+1}(t)}.$$

The following quantities can therefore be substituted into Eq. (II.168):

from 1a) for $p=m=0(\neq 1)$ and $z=v$

$$f_m(v) (=f_p(z) = z^{1-p}) = v^{1-0} = v$$

from 1b) for $p=l=1$ and $z=\Delta x$

$$f_l(\Delta x) (=f_p(z) = \ln z) = \ln \Delta x = \ln (1/k)$$

for $p=l=1$ and $z=\Delta x_{min}$

$$f_l(\Delta x_{min}) (=f_p(z) = \ln z) = \ln \Delta x_{min} = \ln (1/k_{max})$$

from 2b)

$$\beta = \alpha_0 f_l(\Delta x_{min}) = \alpha_0 \ln (1/k_{max}).$$

The results are

$$v = -\alpha_0 \ln \frac{1}{k} + \alpha_0 \ln (1/k_{max})$$

as well as

$$q = vk = k\alpha_0 [\ln (1/k_{max}) - \ln (1/k)].$$

This function has a maximum at which $v=v_{opt}$ and $k=k_{opt}$. The quantity k_{opt} does not denote an optimal density, but only the density corresponding to $v=v_{opt}$. Let us now derive k_{opt} and v_{opt}:

$$dq/dk = (k\alpha_0/k) + \alpha_0 [\ln (1/k_{max}) - \ln (1/k)] = 0$$

$$\ln (1/k_{opt}) = 1 + \ln (1/k_{max})$$

$$k_{opt} = e^{-(1 + \ln 1/k_{max})} = \frac{1}{e} k_{max}$$

$$v_{opt} = \alpha_0 [\ln (1/k_{max}) - \ln (e/k_{max})] = -\alpha_0 \ln e$$

$$\alpha_0 = -v_{opt}.$$

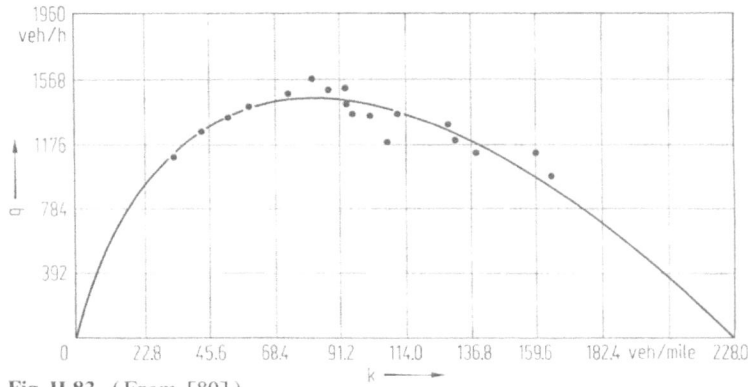

Fig. II.83. (From [80])

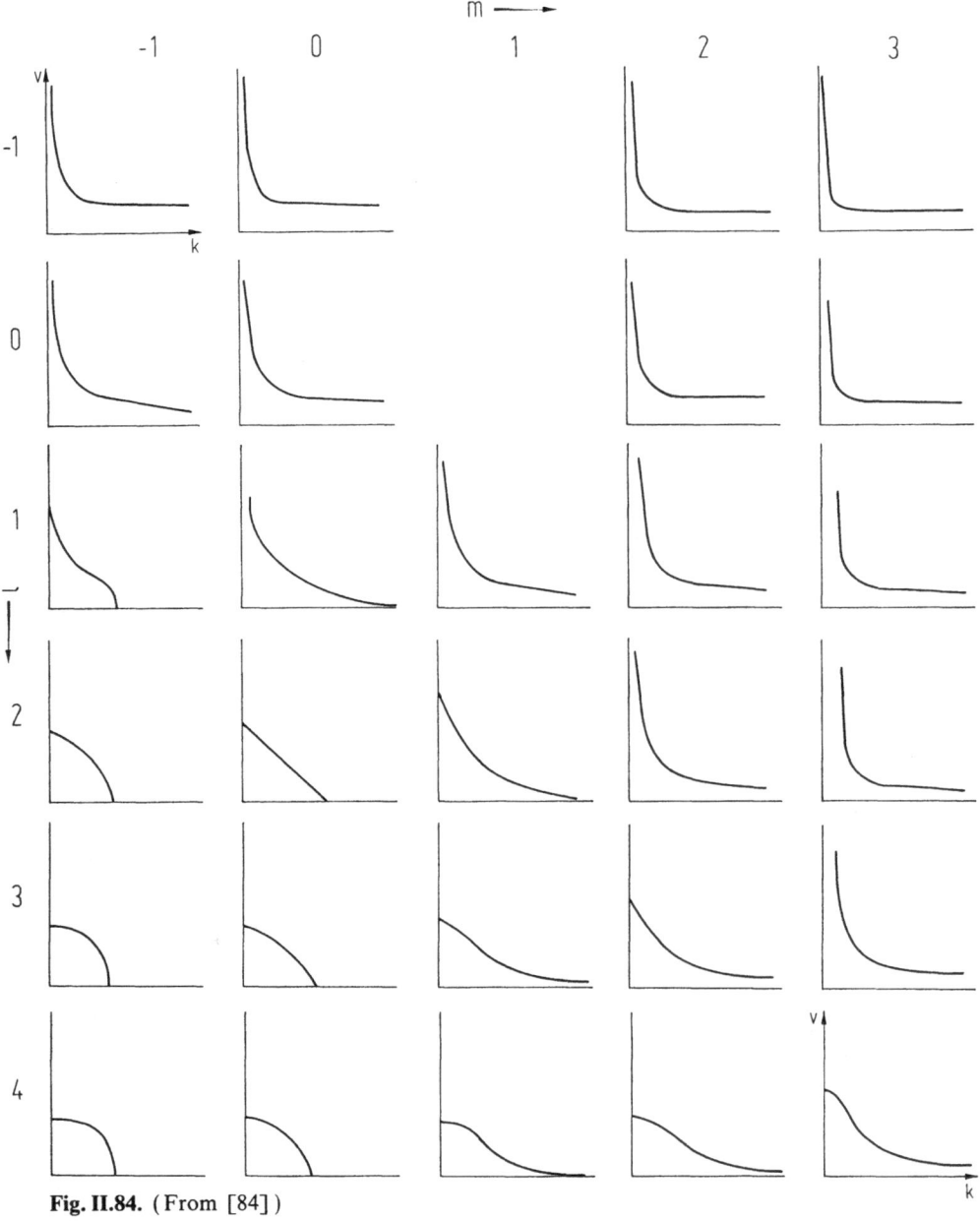

Fig. II.84. (From [84])

Finally,

$$v = -v_{opt}[\ln(1/k_{max}) - \ln(1/k)] = v_{opt}\ln(k_{max}/k) \tag{II.169}$$

and

$$q = vk = v_{opt}k\ln(k_{max}/k). \tag{II.170}$$

This model obviously fits observed data better, as Fig. II.83 shows for data from American investigations.

Consider now the model with $m=0$ and $l=2$:

$$\ddot{x}_{n+1}(t+T) = \alpha_1 \frac{\dot{x}_n(t) - \dot{x}_{n+1}(t)}{[x_n(t) - x_{n+1}(t)]^2}.$$

In this case

1) for $p=m=0$ and $z=v$: $f_m(v) = v^{1-0} = v$
2) for $p=l=2$ and $z=\Delta x$: $f_l(\Delta x) = \Delta x^{1-l} = \Delta x^{-1} = k$
3) $\beta = \alpha_1 f(\Delta x_{min}) = \alpha_1 k_{max}$

we obtain

$$v = \alpha_1 k_{max} - \alpha_1 k.$$

α_1 is determined by the boundary condition that for $k=0$ the speed must be $v=v_w$. Therefore,

$$\alpha_1 = \frac{v_w}{k_{max}}$$

(II.171)

$$v = v_w \frac{k_{max}}{k_{max}} - v_w \frac{k}{k_{max}} = v_w \left(1 - \frac{k}{k_{max}}\right)$$

and then

$$q = vk = v_w k \left(1 - \frac{k}{k_{max}}\right).$$

(II.172)

Figure II.84 shows how different combinations of m and l influence the shape of the function $v=v(k)$ (and of course also $q=q(k)$). Parameters m and l need not be integers. Investigations have shown that a model with $m=0.8$ and $l=2.8$ is a very good fit to data observed on an expressway in Chicago. Because of the shape of the curve $v=v(k)$ compare Fig. II.84 with Figs. II.45 and II.47.

II.3.3.2 Psycho-Physical Spacing Models

The car following equations discussed in Sect. II.3.3.1.3 assume that the driver of the following vehicle reacts, on the one hand, to arbitrarily small changes in the relative speed $\Delta\dot{x}$ and, on the other hand, even at very large spacings. Therefore these equations assume that there is no response as soon as speed differences disappear [see Eq. (II.165)]. These assumptions are not very realistic.

Psycho-physical models make possible a description of such processes which is substantially closer to reality. Research into perceptual psychology has shown that drivers are subject to certain limits on the stimuli to which they respond. The basis of such models is:

1) at large spacings the driver of a following vehicle is not influenced by the size of the speed difference, and
2) at small spacings there are combinations of relative speeds and distance-headways for which there is, as in 1), no response of the driver of the following vehicle because the relative motion is too small.

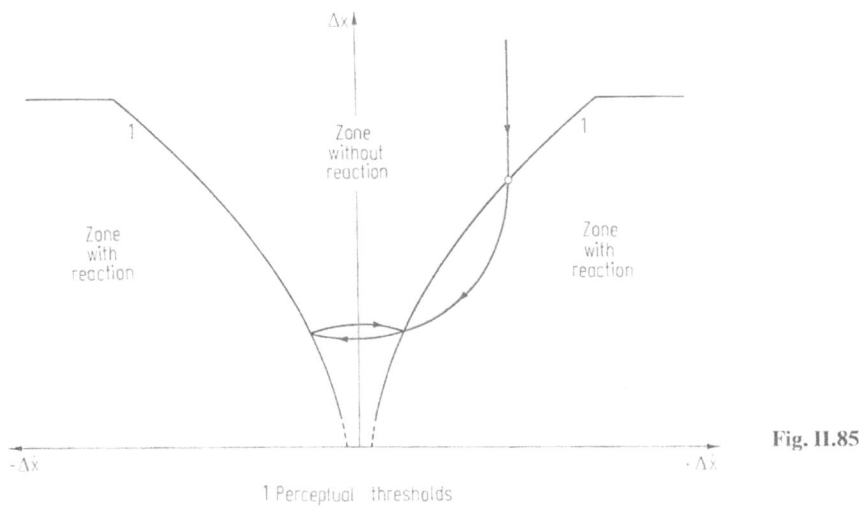

Fig. II.85

1 Perceptual thresholds

This means that there are perceptual thresholds. Only when these thresholds are reached will the driver of a following vehicle be able to perceive the change in the apparent size of the leading vehicle and subsequently be able to react to the changes in the kinematic variables. Such thresholds are represented as parabolas in a Δx-$\Delta \dot{x}$-coordinate system (Fig. II.85).

It can also be seen from this picture how car following proceeds. A vehicle with speed $\dot{x}_{n+1}$, which is larger than the speed $\dot{x}_n$ of the preceding vehicle will catch up with constant relative speed $\Delta \dot{x}$. Upon reaching the threshold the driver reacts by reducing his speed. One such example of relative motion with constant deceleration appears as a parabola. The minimum of the parabola lies on the Δx-axis. The driver tries to decelerate so as to reach a point at which $\Delta \dot{x} = 0$. He is not able to do this accurately because, first, he is not able to perceive small speed differences and, second, he is not able to control his speed sufficiently well. The result is that the spacing will again increase. When the driver first reaches the opposite threshold he accelerates and tries again to achieve the desired spacing (indicated in Fig. II.85 by the upper part of the loop).

If one assumes that the relationship of the perceptual thresholds for spacing are the same for both positive and negative changes in relative speed, then the resulting spacing behaviour resembles a symmetrical pendulum about its equilibrium point.

In fact it has been observed that the perceptual threshold of a single "test subject" shows, in addition, random variations in the location of the threshold in a Δx-$\Delta \dot{x}$-coordinate system (Fig. II.86). Furthermore, the perceptual threshold for positive speed differences (the spacing decreases) is smaller than that for negative speed differences (the spacing increases). One plausible explanation for this behaviour is that drivers pay closer attention to spacing decreases than to spacing increases simply on the basis of their own safety. As Fig. II.86 shows, the resulting behaviour is a cyclical variation superposed on an increasing mean spacing. As can be verified by observation, this so-called "drift" is compensated for, by the

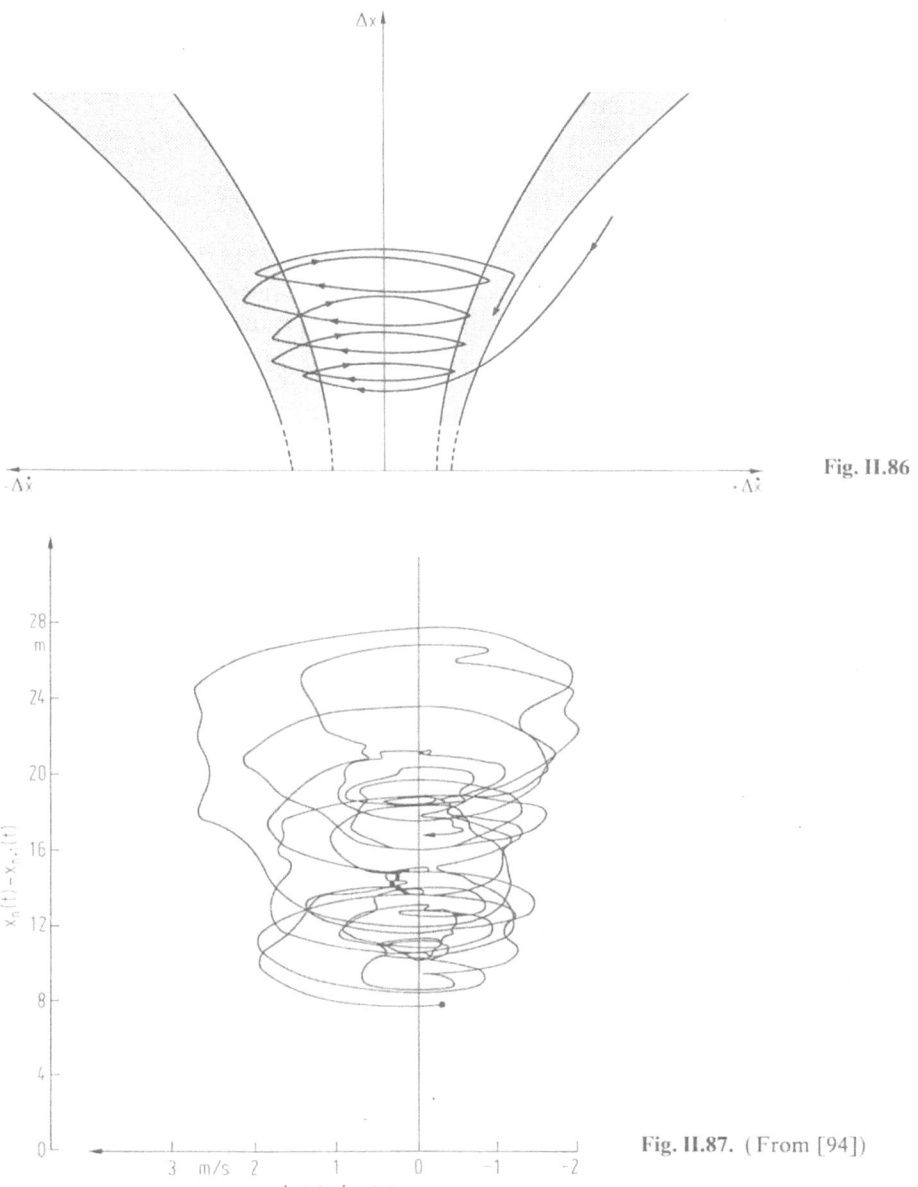

Fig. II.86

Fig. II.87. (From [94])

driver, in two different ways: (1) if a driver is at some point moving faster than the lead car, he ceases his positive acceleration and travels instead at constant speed; or (2) he assumes a different acceleration in the positive quadrant of the $\Delta\dot{x}$-Δx-coordinate system than in the negative quadrant.

As Fig. II. 87 shows, these distance oscillations and the bunching effect are also observed in actual traffic. Psycho-physical spacing models therefore include

several substantially more realistic assumptions concerning traffic flow phenomena. They are better suited than deterministic spacing models for simulation
models. It can, however, be shown that the behaviour described corresponds to the
general car following equation when m = 0 and l = 2.

II.3.3.3 Continuum Theory

II.3.3.3.1 The Fundamental Equation and the Continuum Equation

In Sects. II.2.1 and II.2.2., the quantities k and q were defined for discrete flows.
That is, when one makes a local measurement at position x_i over the time interval
$t = t_i - t_0$ one obtains [cf. Eq. (II.15)]

$$q = \frac{\Phi_{x_i}(t_i) - \Phi_{x_i}(t_0)}{\Delta t} = \frac{M(x_i, t, \Delta t)}{\Delta t},$$

and, further, when one makes two local measurements at positions x_0 and
$x_i = x_0 + \Delta x$ one obtains for time t_i [see Eq. (II.18)]

$$k = -\frac{\Phi_{x_i}(t_i) - \Phi_{x_0}(t_i)}{\Delta x} = \frac{N(t_i, x, \Delta x)}{\Delta x}.$$

Taking the limits, one finds, respectively [see Eqs. (II.16) and (II.19)]

$$\lim_{\Delta t \to 0} \frac{P[M(x_i, t, \Delta t) \geqq 1]}{\Delta t} = \lambda_x(t)$$

$$\lim_{\Delta x \to 0} \frac{P[N(t_i, x, \Delta x) \geqq 1]}{\Delta x} = \varkappa_t(x).$$

These limits refer to the discrete flow which was represented in Fig. II.11 as a three-
dimensional staircase in the x-t plane. It follows then that if n(x,t) is taken to be a
continuous plane, then $\lambda_x(t)$ and $\varkappa_t(x)$ can be obtained directly as partial
derivations in the t- and x-directions respectively (Fig. II.88). In this way the
transition from a discrete flow process to a continuum flow process is frequently
derived. Furthermore, it is clear that the continuum approximation to a discrete
flow is valid only in the regime of dense traffic.

With these assumptions,

$$\lambda_x(t) = \frac{\delta n(x,t)}{\delta t} \qquad\qquad\qquad (II.173)$$

and

$$\varkappa_t(x) = -\frac{\delta n(x,t)}{\delta x}. \qquad\qquad\qquad (II.174)$$

The total differential of the function n(x,t) is

$$dn(x,t) = \frac{\delta n(x,t)}{\delta t} dt + \frac{\delta n(x,t)}{\delta x} dx$$

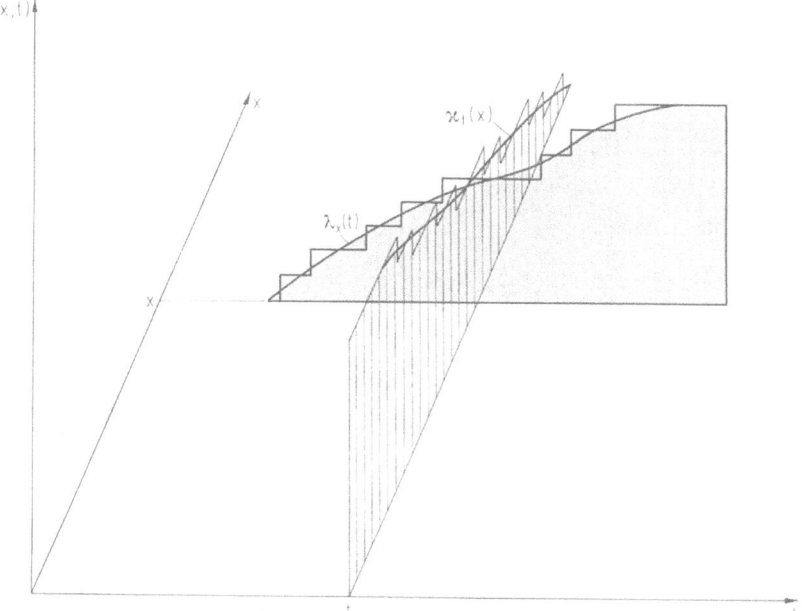

Fig. II.88

so that the total derivative with respect to t is

$$\frac{dn(x,t)}{dt} = \frac{\delta n(x,t)}{\delta t} + \frac{\delta n(x,t)}{\delta x}\frac{dx}{dt}.$$

Setting this derivative to zero, using Eqs. (II.173) and (II.174), and noting that $dx/dt = v$ we obtain, by analogy with Eq. (II.57),

$$\lambda_x(t) - \varkappa_t(x)v = 0$$

$$\lambda_x(t) = v\varkappa_t(x). \tag{II.175}$$

If one begins instead with instantaneous measurements (working with n' instead of n) Eq. (II.21) states that

$$q = -\frac{n'_{t_i}(x_i) - n'_{t_0}(x_i)}{\Delta t}$$

from which

$$\lambda_x(t) = -\frac{\delta n'(x,t)}{\delta t}. \tag{II.176}$$

Similarly, Eq. (II.20)

$$k = \frac{n'_{t_i}(x_i) - n'_{t_i}(x_0)}{\Delta x}$$

yields the result

$$\varkappa_t(x) = \frac{\delta n'(x,t)}{\delta x}. \tag{II.177}$$

The total derivative of $n'(x,t)$ with respect to x, set equal to zero, gives

$$\frac{dn'(x,t)}{dx} = \frac{\delta n'(x,t)}{\delta t} \frac{dt}{dx} + \frac{\delta n'(x,t)}{\delta x} = 0$$

and, noting that $dt/dx = w$,

$$\varkappa_t(x) = w\lambda_x(t) \tag{II.178}$$

by analogy with Eq. (II.58). Equations (II.175) and (II.178) are referred to as the fundamental equations of continuum theory. Setting $\varkappa = \varkappa(x,t)$, its total derivative with respect to time is

$$\frac{d\varkappa(x,t)}{dt} = \frac{\delta\varkappa(x,t)}{\delta t} + \frac{\delta\varkappa(x,t)}{\delta x} \frac{dx}{dt}.$$

If $d\varkappa(x,t)/dt$ is set to zero and substituting

$$\varkappa(x,t) = -\frac{\delta n(x,t)}{\delta x}$$

we obtain

$$\frac{d\varkappa(x,t)}{dt} = -\frac{\delta^2 n(x,t)}{\delta x \delta t} - \frac{dx}{dt}\frac{\delta^2 n(x,t)}{\delta x^2} = 0. \tag{II.179}$$

Analogously, we can write

$$\frac{d\lambda(x,t)}{dx} = \frac{\delta\lambda(x,t)}{\delta x} + \frac{\delta\lambda(x,t)}{\delta t}\frac{dt}{dx} = 0. \tag{II.180}$$

Substitute

$$\lambda(x,t) = -\frac{\delta n'(x,t)}{\delta t}$$

we obtain

$$\frac{d\lambda(x,t)}{dx} = -\frac{\delta^2 n'(x,t)}{\delta t \delta x} - \frac{dt}{dx}\frac{\delta^2 n'(x,t)}{\delta t^2} = 0. \tag{II.181}$$

These two partial differential equations are equations of continuity. Now let us summarize the results derived so far:

$$\frac{\delta n'(x,t)}{\delta t} + \frac{dx}{dt}\frac{\delta n'(x,t)}{\delta x} = 0$$

$$\frac{\delta n'(x,t)}{\delta t}\frac{dt}{dx} + \frac{\delta n'(x,t)}{\delta x} = 0$$

from which the following continuity equations are derived

$$\frac{\delta^2 n(x,t)}{\delta x \delta t} + \frac{dx}{dt} \frac{\delta^2 n(x,t)}{\delta x^2} = 0$$

$$\frac{\delta^2 n'(x,t)}{\delta t \delta x} + \frac{dt}{dx} \frac{\delta^2 n'(x,t)}{\delta t^2} = 0.$$

One example will illustrate the use of the function $n(x,t)$:

Example 40. Let us imagine a traffic stream in which all vehicles travel at the same constant speed v but where at a particular point $x_0 = 0$ there is instationarity of the form

$$n(0,t) = a \cdot t^2, \quad a > 0.$$

Then, in general,

$$n(x,t) = a \left(t - \frac{x}{v} \right)^2.$$

From this we obtain

$$\lambda(x,t) = \frac{\partial n(x,t)}{\partial t} = 2a \left(t - \frac{x}{v} \right).$$

A cross-section at right angles to the time axis at t_i yields

$$n(x,t_i) = a \left(t_i - \frac{x}{v} \right)^2$$

which in turn yields

$$\varkappa(x,t_i) = -\frac{\partial n(x,t_i)}{\partial x} = \frac{2a}{v} \cdot \left(t_i - \frac{x}{v} \right).$$

Consequently, concentration is also not stationary over distance.

Variable v is obtained when a cross-section is made at high $n(x,t) = \text{const} = c$ parallel to the x-t plane. Then (for control)

$$n(x,t) = a \left(t - \frac{x}{v} \right)^2 = c, \quad c \geq 0$$

and from this

$$x(t) = v(t \pm \sqrt{c/a}); \quad dx/dt = v = \text{const}$$

as assumed.

The continuity equations can also be derived by means of probability theory or by using the function $n'(x,t)$. The probability that a vehicle is located in dx at time t is $\varkappa(x,t)dx$ (see Sects.II.2.1 and II.2.2). The probability that a vehicle enters the distance interval dx during the time interval dt is $\lambda(x,t)dt$. Two equivalent events

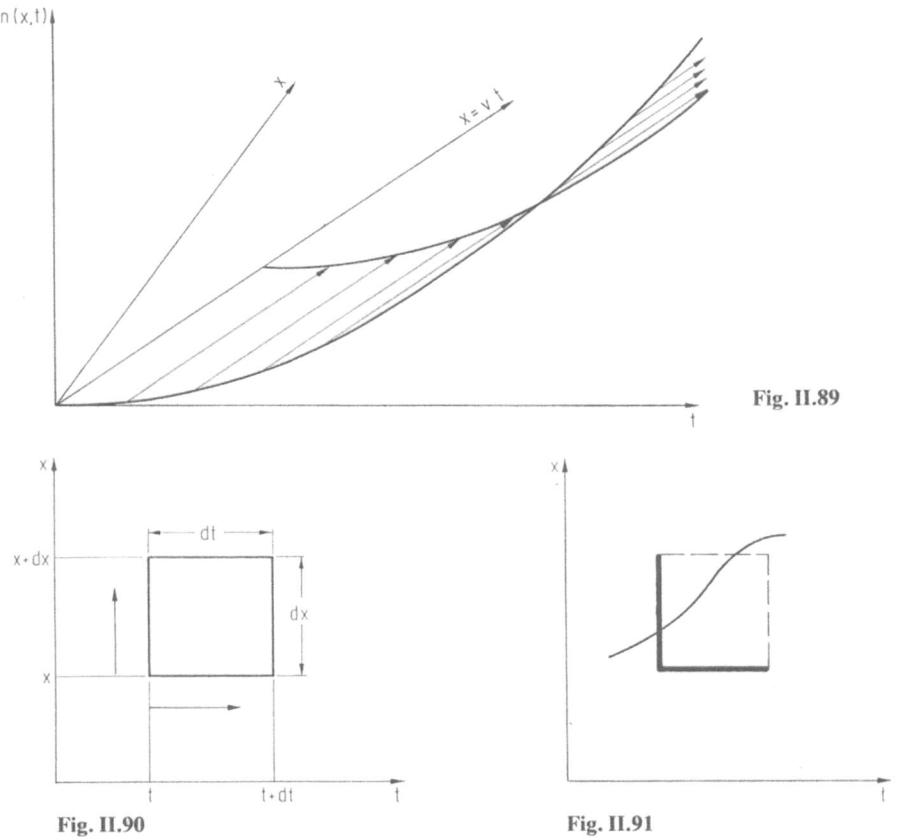

Fig. II.89

Fig. II.90   Fig. II.91

which therefore have the same probability can be defined:

(A) A vehicle either enters the distance interval dx during the time interval dt (event A_1) or is located in dx at time t (event A_2).

(B) The same vehicle is either located in dx at time $t + dt$ (event B_1) or leaves dx during the time interval dt (event B_2). Those trajectories which cross the more heavily drawn sides of the delineated $\Delta x - \Delta t$ element must also cross the dashed sides of the $\Delta x - \Delta t$ element (see Fig. II.91).

Since events A_1 and A_2 are mutually exclusive, and events B_1 and B_2 are also mutually exclusive, the preceding statement that events A and B are equivalent results in the equation[1]

$$\varkappa(x,t)\,dx + \lambda(x,t)\,dt = \varkappa(x,t+dt)\,dx + \lambda(x+dx,t)\,dt$$

or

$$\varkappa(x,t)\,dx - \varkappa(x,t+dt) + \lambda(x,t)\,dt - \lambda(x+dx,t)\,dt = 0.$$

1 For mutually exclusive events A_1 and A_2, $P(A_1+A_2) = P(A_1) + P(A_2) = P(A)$ and similarly for B_1 and B_2.

With

$$\varkappa(x,t)\,dx - \varkappa(x,t+dt)\,dx = [\varkappa(x,t) - \varkappa(x,t+dt)]\,dx = \frac{\delta\varkappa(x,t)}{\delta t}\,dt\,dx$$

and

$$\lambda(x,t)\,dt - \lambda(x+dx,t)\,dt = [\lambda(x,t) - \lambda(x+dt,t)]\,dt = \frac{\delta\lambda(x,t)}{\delta x}\,dx\,dt$$

Eq. (II.182) can be written

$$\frac{\delta\varkappa(x,t)}{\delta t}\,dt\,dx + \frac{\delta\lambda(x,t)}{\delta x}\,dx\,dt = 0$$

thereby resulting in the continuity equation

$$\frac{\delta\varkappa(x,t)}{\delta t} + \frac{\delta\lambda(x,t)}{\delta x} = 0. \tag{II.183}$$

Since the intensity is a function of the concentration

$$\lambda = \lambda(\varkappa,x,t)$$

we have

$$\frac{\delta\lambda(\varkappa,x,t)}{\delta t} = \frac{\delta\lambda}{\delta\varkappa}\,\frac{\delta\varkappa(x,t)}{\delta x}$$

so that

$$\frac{\delta\varkappa(x,t)}{\delta t} + \frac{\delta\lambda}{\delta\varkappa}\,\frac{\delta\varkappa(x,t)}{\delta x} = 0. \tag{II.184}$$

Comparing Eq. (II.184) with the preceding Eq. (II.179) it is shown that

$$\frac{\delta\lambda}{\delta\varkappa} = \frac{dx}{dt} = c. \tag{II.185}$$

II.3.3.3.2 Shockwaves

When the traffic stream is stationary over time and distance, then $\varkappa$ and v (and therefore also λ) are independent of x and t (see also Sect. II.3.1.1). Such a traffic stream is described mathematically by a plane in n-x-t-space:

$$n(x,t) = \lambda t - \varkappa x + a$$

with $\delta n(x,t)/\delta t = \lambda = \text{const}$, $\delta n(x,t)/\delta x = -\varkappa = \text{const}$, and $v = \lambda/\varkappa = \text{const}$.

When the traffic state changes from $(\lambda_1,\varkappa_1)$ to $(\lambda_2,\varkappa_2)$ (see Fig. II.79), the location of this change in state can be represented by the intersection of the two planes

$$n_1 = \lambda_1 t - \varkappa_1 x + a_1$$

$$n_2 = \lambda_2 t - \varkappa_2 x + a_2$$

(see Sect. II.3.3.1.4).

Along this intersection

$$\frac{\delta n_1}{\delta t} = \frac{\delta n_2}{\delta t}$$

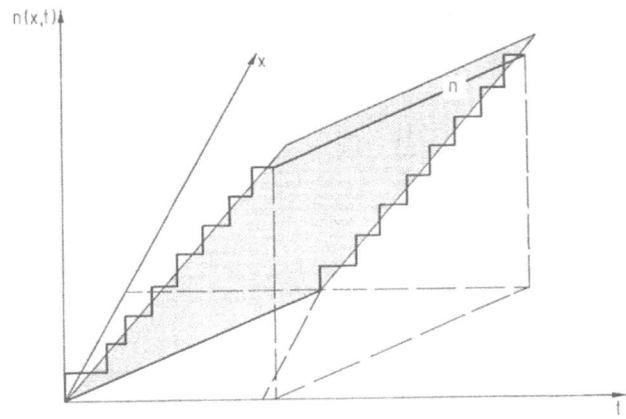

Fig. II.92

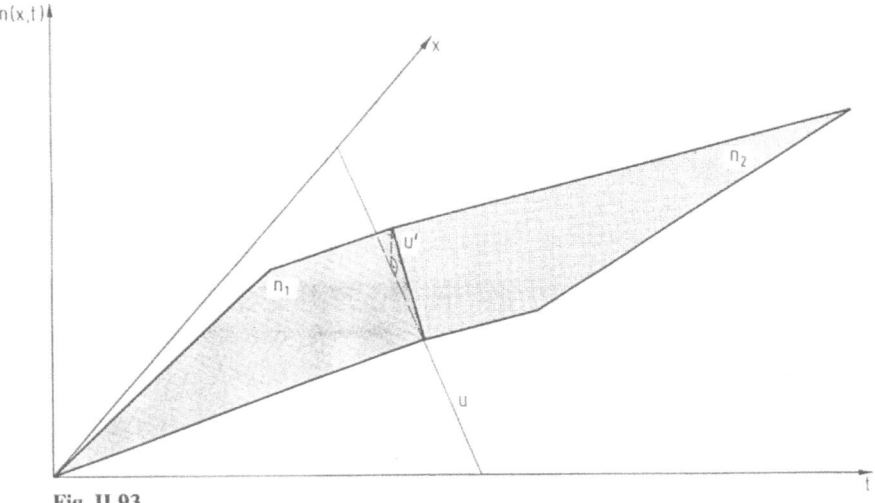

Fig. II.93

or

$$\lambda_1 - \varkappa_1 \frac{dx}{dt} = \lambda_2 - \varkappa_2 \frac{dx}{dt},$$

from which the slope of the projection of this intersection on to the x-t plane can be calculated. This slope is the propagation speed of the change of state (commonly referred to as a shockwave) and is written as

$$\frac{dx}{dt} = u = \frac{\lambda_1 - \lambda_2}{\varkappa_1 - \varkappa_2} = \frac{\Delta\lambda}{\Delta\varkappa} \tag{II.186}$$

(see Fig. II. 93).

Such a shockwave moves with speed u, such that $\lambda_1 - u\varkappa_1$ vehicles per unit time enter it and $\lambda_2 - u\varkappa_2$ vehicles per unit time leave it. This can be illustrated as

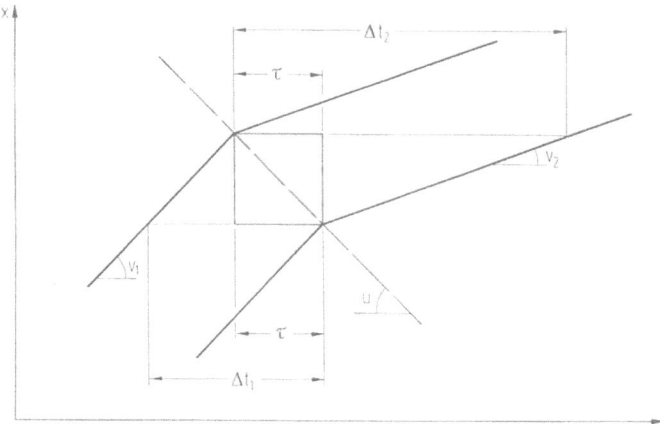

Fig. II.94

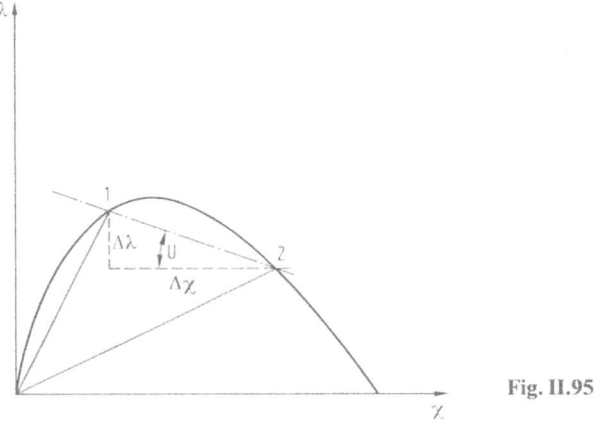

Fig. II.95

follows: If an observer stands at location x, then he counts $\lambda = v\varkappa$ vehicles per unit-time. If, on the other hand, the observer moves with a speed u in the direction of the traffic stream, then he counts only $(v-u)\varkappa = \lambda - u\varkappa$ vehicles per unit time, where a negative value of u means that the observer is travelling against the traffic stream. Therefore, the average headway with which vehicles traverse the shockwave, and thereby change speed and distance, is

$$\tau = \frac{1}{\lambda_1 - u\varkappa_1} = \frac{1}{\lambda_2 - u\varkappa_2}$$

(see Fig. II.94). From this relationship Eq. (II.186) can also be derived.

For state (λ_1, τ_1), $v_1 = \lambda_1/\tau_1$ is assumed constant, $\Delta t_1 = 1/\lambda_1$ is the time-headway and $\Delta x_1 = 1/\varkappa_1$ the distance between two vehicles. Thus for every point in the fundamental diagram there is a defined set of parallel straight lines (lines of motion) in the x-t plane. The transition (the speed change) from one state to the other takes place along the shockwave lines. The two states $(\lambda_1, \varkappa_1)$ and $(\lambda_2, \varkappa_2)$ are two points on the fundamental diagram; it is easy to show that u is simply the slope of the cord connecting these two points (Fig. II.95).

By choice of appropriate units, the speeds v_i of the invididual vehicles [equal to the radius vector to the point $(\lambda_1, \varkappa_1)$ in the fundamental diagram] and the speed of the shockwave are parallel in both diagrams.

Example 41. The traffic flow on a motorway is $q_1 = 2\,000$ veh/h with $\bar{v}_{m_1} = 80$ km/h. As the result of an accident, the motorway is blocked. The density in the queue is $k_2 = 275$ veh/km (on both of the lanes).

a) At which rate does the queue increase? The answer is

$$k_1 = \frac{q}{\bar{v}_{m_1}} = \frac{2000}{80} = 25 \text{ veh/km}$$

$$\bar{v}_{m_2} = 0; \quad q_2 = k_2 \bar{v}_{m_2} = 0$$

$$u = \frac{2000 - 0}{25 - 275} = -8 \text{ km/h}$$

so that — obviously — the queue grows in the direction against traffic.

b) What is the rate at which the queue grows, in units of vehicles per hour? The queue grows at

$$q_1 - k_1 u = q_2 - k_2 u$$

$$2000 - (-25.8) = 0 - (-275.8) = 2200 \text{ veh/h}.$$

The growth of a queue can be represented graphically. As mentioned previously an observer moving with the shockwave sees a traffic flow

$$\lambda' = \varkappa_2 (v_2 - u) \quad [= \varkappa_1 (v_1 - u)].$$

The total stoppage results in an output state with $v_2 = 0$ and $\varkappa_2 = \varkappa_{max}$, so that

$$\lambda' = \varkappa_{max} u \quad \text{and} \quad u = \lambda'/k_{max}.$$

Figure II.96b shows λ' as the intersection on the ordinate of the extension of a straight line connecting the two states $(\lambda_1, \varkappa_1)$ and $(\lambda_2, \varkappa_2)$.

Assuming the input flow to be stationary over time, the number of vehicles entering the queue in Δt is $N = \lambda' \Delta t$. For the graphical representation of the growth of the queue, the ordinate λ' must be transformed into a slope. This procedure is shown in the left hand quadrant of Fig. II.96b. Once the straight line having slope λ' has been transformed as in Fig. II.96a, it is further possible to construct a distance scale. Transforming $\varkappa_{max}$ to a slope, as shown in Fig. II.96b, and transforming the resulting straight line to the left quadrant of Fig. II.96a, it is possible to determine the length Δx required for a queue of N vehicles which builds up during time Δx. Of course, one can alternatively determine the time Δt required for a queue to grow to a length Δx. As a further graphical refinement, the distance scale can be transferred to the ordinate, as shown in the upper quadrant of Fig. II.97.

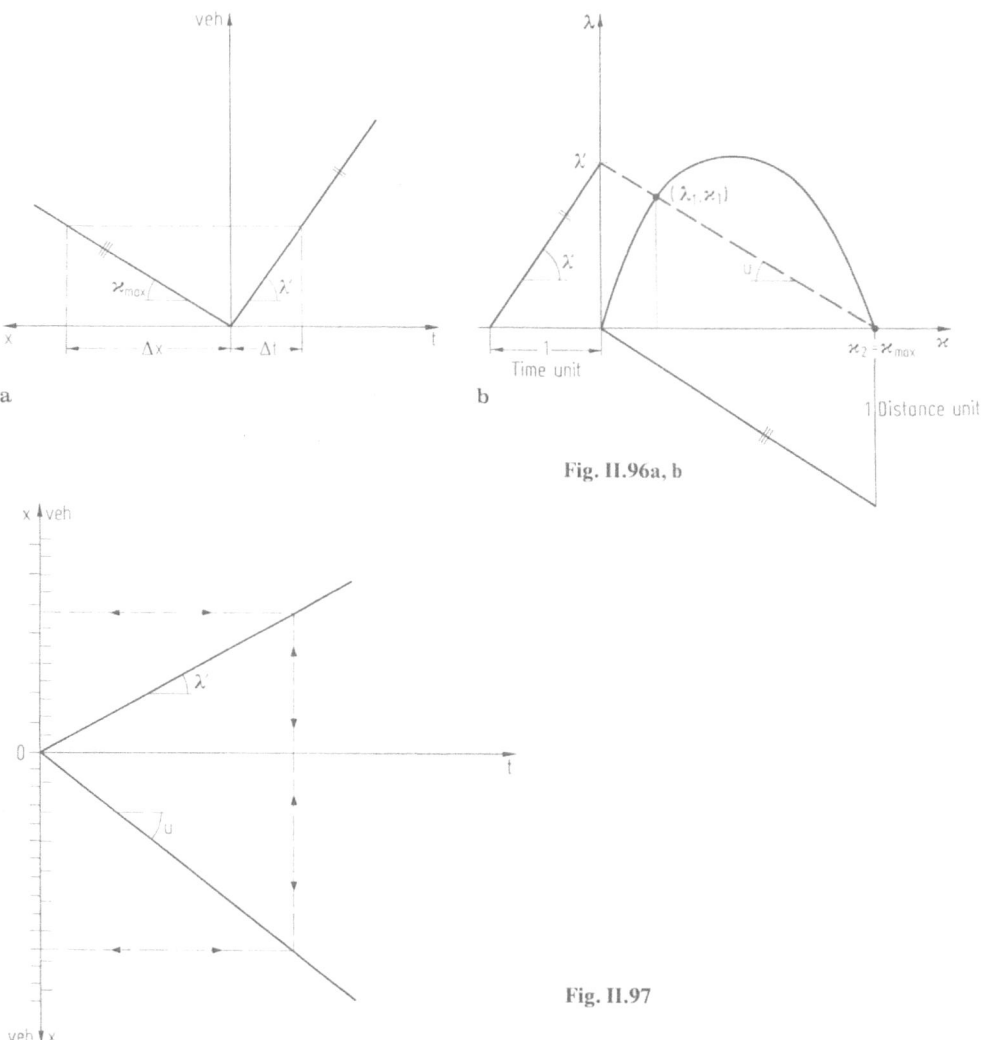

Fig. II.96a, b

Fig. II.97

Figure II.98 represents the state queue in three dimensions. Figure II.97 also includes a compact representation of Fig. II.98, showing the growth of the queue in terms of u directly.

In general, the input is not stationary over time. So long as the fundamental diagram is assumed to be invariant with respect to distance, it is generally an acceptable approximation to assume piecewise stationarity of the input over time, so that graphical methods remain applicable. Figure II.99 illustrates an example.

Now let the cause of the flow blockage be removed after a time T. Let it be assumed that the output flow has the value λ_3 (e.g. $\lambda_3 = \lambda_{max}$, the assumption that $\lambda_3 = \lambda_{max}$ will not be explored further), so that

$$u' = \lambda_3 / (\varkappa_3 - \varkappa_{max})$$

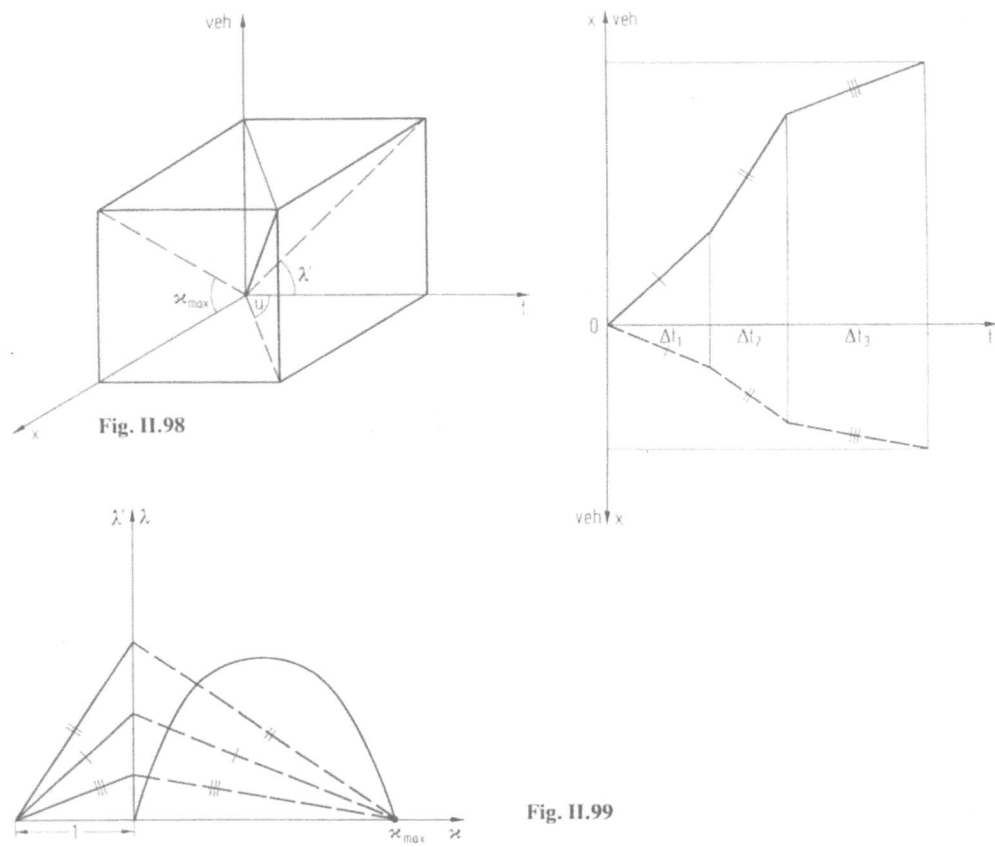

Fig. II.98

Fig. II.99

by which the discharge of the queue can likewise be graphically represented. Figure II.100 shows an example in which λ_1 and λ_3 are stationary, and $\lambda_3 > \lambda_1$. The vehicle trajectories are drawn assuming instantaneous changes in speed, in conformance with the underlying theory. Furthermore it should be noted that a shockwave remains even after the queue has dissipated, until the vehicles from the queue (contrary to the assumption in the example) have accelerated from v_3 to v_1. This acceleration is suggested by the dashed trajectory for the first vehicle.

Example 42. On a section of a motorway, an accident occurs at 10:00 am at point B (see Fig. II.101). At first, the vehicles involved block the entire carriageway. After 15 min, one lane is cleared, and thereafter one lane of traffic flows past point B.
Data:

Flow at A:	$q = 2700$ veh/h,	$\bar{v}_m = 90$ km/h
Flow at B, one-lane:	$q = 1500$ veh/h,	$\bar{v}_m = 7.5$ km/h
Flow at B, two-lane:	$q = 3600$ veh/h,	$\bar{v}_m = 60$ km/h
Queue, density:	$k_{max} = 300$ veh/km.	

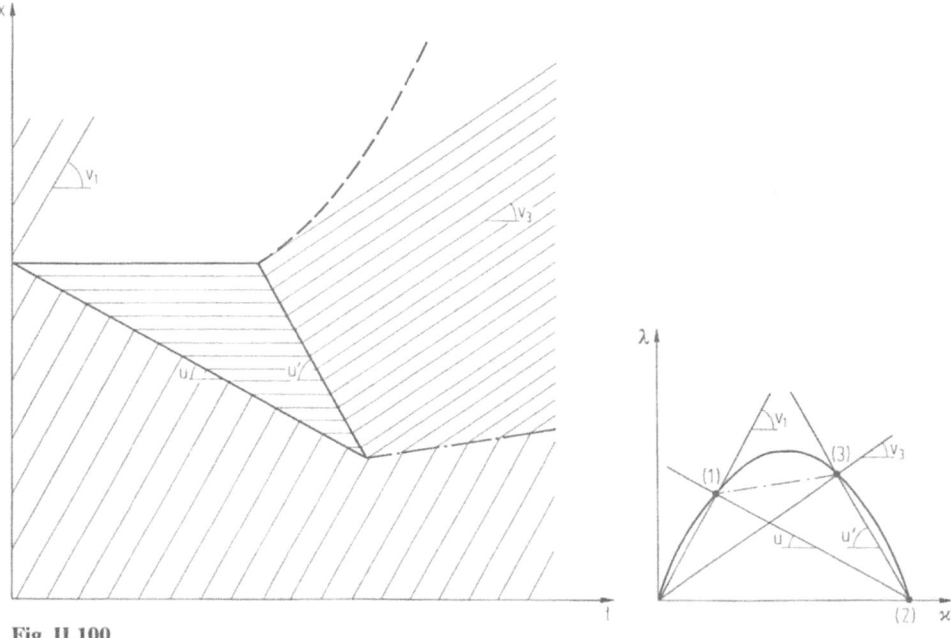

Fig. II.100

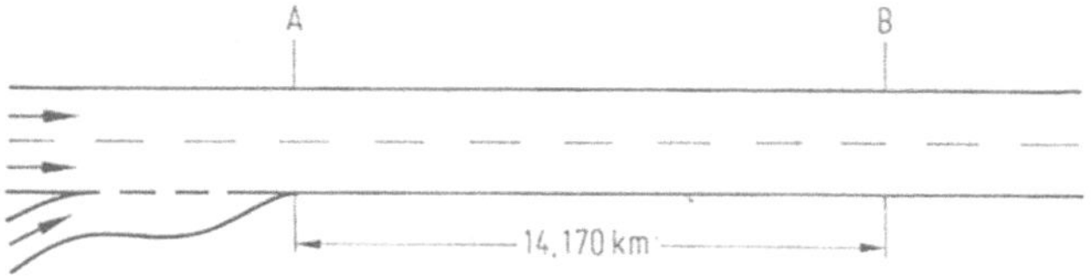

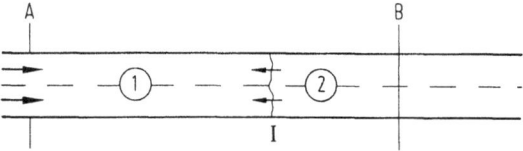

Fig. II.101

─ 14,170 km ─

Answer the following questions:

1. Where is the end of the queue at 10:15 am?
2. When are vehicles last forced to stop by the queue?
3. What is the maximum queue size?
4. What is the maximum distance of the end of the queue from the site of the accident?
5. By what time must the second lane be cleared if the disturbance to the traffic flow resulting from the accident is not to extend to entrance A?

Concerning 1: The situation is illustrated in Fig. II.102.

Fig. II.102. 1, region of undisturbed traffic; 2, queue (stationary vehicles); I, shockwave

The speed of the shockwave is given by

$$u_I = \frac{q_1 - q_2}{k_1 - k_2}.$$

Given $k_1 = q_1/\bar{v}_{m_1} = 2700/90 = 30$ veh/km and $k_2 = k_{max} = 300$ veh/km and $q_2 = 0$ veh/h (queue) we obtain

$$u_I = \frac{2700 - 0}{30 - 300} = -10 \text{ km/h}.$$

This indicates that after 15 min the end of the queue has covered a distance of

$$x_I(t = 15 \text{ min}) = u_I \cdot \frac{15}{60} = -2.5 \text{ km}.$$

Concerning 2: The corresponding situation is shown in Fig. II.103.

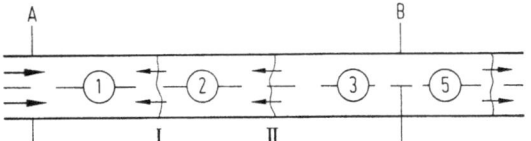

Fig. II.103

After the partial clearing of the blockage resulting from the accident, traffic flows at $q_3 = 1500$ veh/h. A shockwave II forms at the barrier between regions 2 and 3 (namely the front of the queue) which follows shockwave I. The last vehicle stops when the front of the queue meets the back of the queue [namely when $x_I(t^*) = x_{II}(t^*)$].

The speed of shockwave II is obtained as follows

$$u_{II} = \frac{q_2 - q_3}{k_2 - k_3}.$$

Given $k_3 = q_3/\bar{v}_{m_3} = 1500/7.5 = 200$ veh/km we obtain

$$u_{II} = \frac{0 - 1500}{300 - 200} = -15 \text{ km/h}.$$

The equations of motion of the two shockwaves are

$$x_I = x(t = 15) + u_I \cdot t = -2.5 - 10 \cdot t$$
$$x_{II} = u_{II} \cdot t = -15 \cdot t.$$

For t^* we obtain

$$-2.5 - 10 \cdot t^* \stackrel{!}{=} -15 \cdot t^*$$

$$t^* = 2.5/(15 - 10)$$

$$t^* = 0.5 \text{ h} = 30 \text{ min}.$$

This indicates that at 10:45 am (30 min after the start of shockwave II) the queue has dissolved.

Concerning 3: The queue reaches its maximum length by the time the shockwave II commences, namely at 10:15 am. At this time, the queue is 2.5 km long (see question 1).

Concerning 4: The maximum distance of the queue from the accident is achieved at 10:45 am (see question 2). This is calculated as follows

$$x_I(t^*) = -2.5 - 10 \cdot 0.5 = -7.5 \, km$$

or alternatively as follows

$$x_{II}(t^*) = -15 \cdot 0.5 = -7.5 \, km.$$

Concerning 5: The time t^{**} at which the disturbance reaches junction A is to be calculated. At 10:45 am shockwaves I and II meet and form a new shockwave which travels with the speed

$$u_{III} = \frac{q_1 - q_3}{k_1 - k_3} = \frac{2700 - 1500}{30 - 200} = -7.06 \, km/h.$$

Time t^{**} is given by

$$t^{**} = t(x = 14.17).$$

From $x(t) = x_{III}(t) = u_{III} \cdot t + x_0 = -7.06 \cdot t - 7.5$ we obtain

$$t^{**} = \frac{14.17 - 7.5}{7.06} = 0.95 \, h = 57 \, min.$$

This indicates that at 11:42 am the disturbance should have reached point A. After the complete removal of the blockage resulting from the accident at point B, a shockwave IV is generated which pursues III, which in turn reaches point A at 11:42 am at the latest.
The situation is depicted in Fig. II.104.

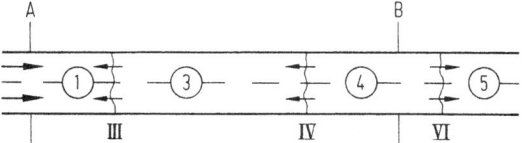

Fig. II.104

The speed of shockwave IV is calculated as follows

$$u_{IV} = \frac{q_3 - q_4}{k_3 - k_4}.$$

Given $k_4 = q_4/\bar{v}_{m4} = 3600/60 = 60 \, veh/km$ we obtain

$$u_{IV} = \frac{1500 - 3600}{200 - 60} = -15 \, km/h.$$

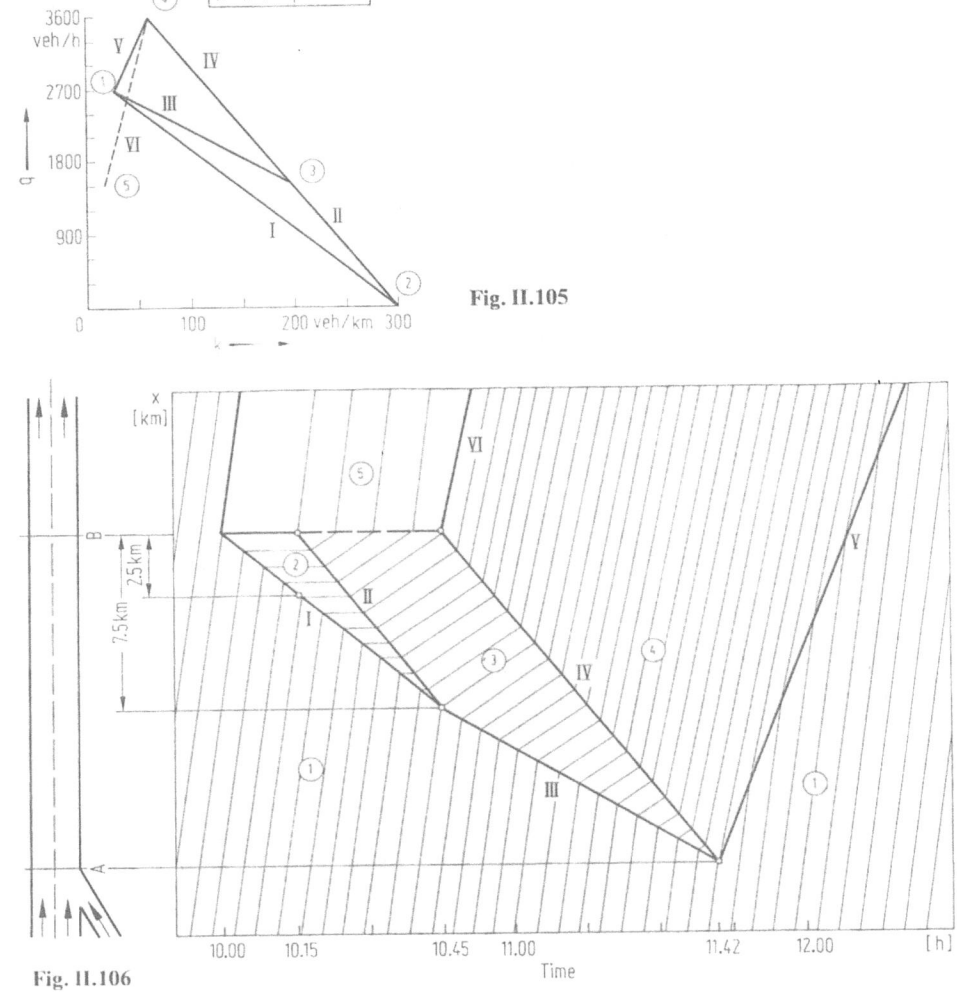

Fig. II.105

Fig. II.106

This indicates that shockwave IV requires

$$\Delta t = \frac{14.17}{15} = 0.95 \text{ h} = 57 \text{ min}$$

for the stretch AB (with length 14.17 km).

The shock wave must therefore commence at 10:45 am at the latest at point B.

Remark: At point A a further shockwave V is formed by the meeting of shockwaves III and IV. This travels downstream with speed given by

$$u_V = \frac{2700 - 3600}{30 - 60} = 30 \text{ km/h}.$$

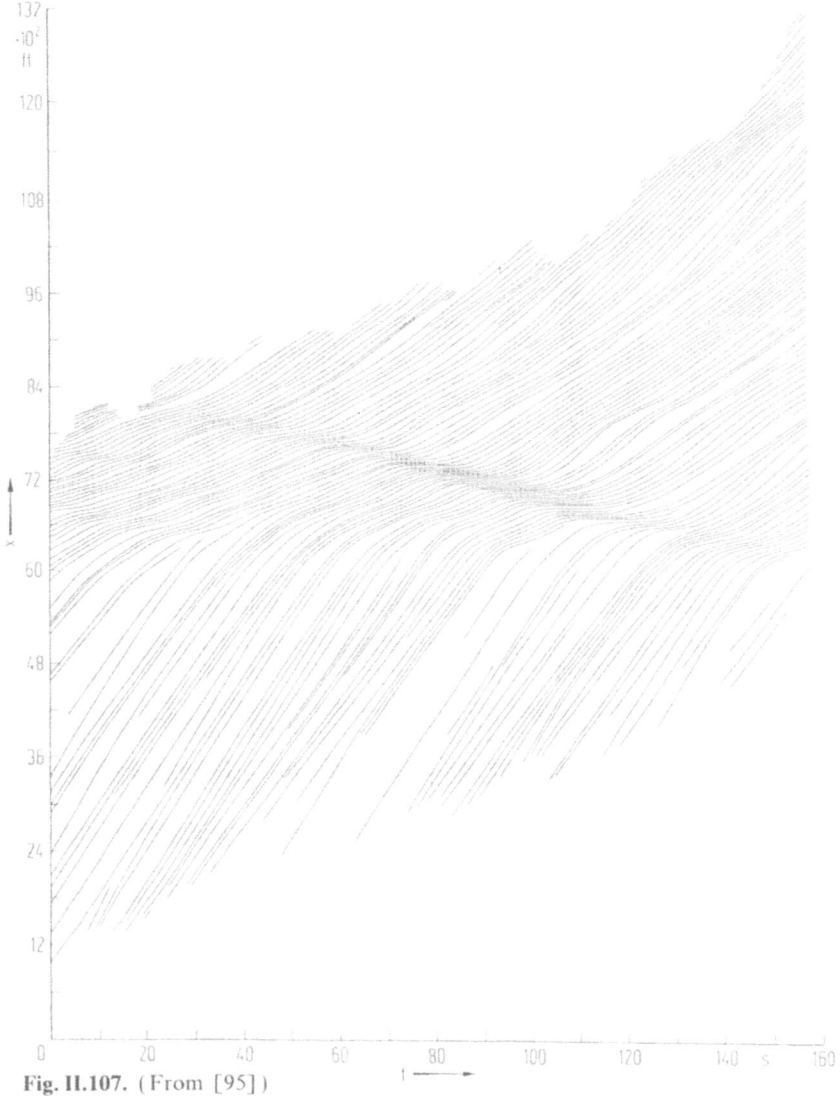

Fig. II.107. (From [95])

At the site of the accident (B), a shockwave VII is created by the clearing of the blockage from the second lane. This travels at high speed, also downstream. Since $u_{VI} > u_V$, these shockwaves do not catch each other up.

In Figs. II.105 and II.106, the different shockwaves and traffic conditions are shown in fundamental diagram and time-distance diagram forms. To determine the speeds of the shockwaves graphically, it is sufficient to enter the corresponding traffic flow conditions as points in the fundamental diagram, without needing to know its entire form.

Figure II.107, constructed from observation of an American freeway, shows the build-up of a traffic jam, having no apparent cause, and its subsequent dissipation.

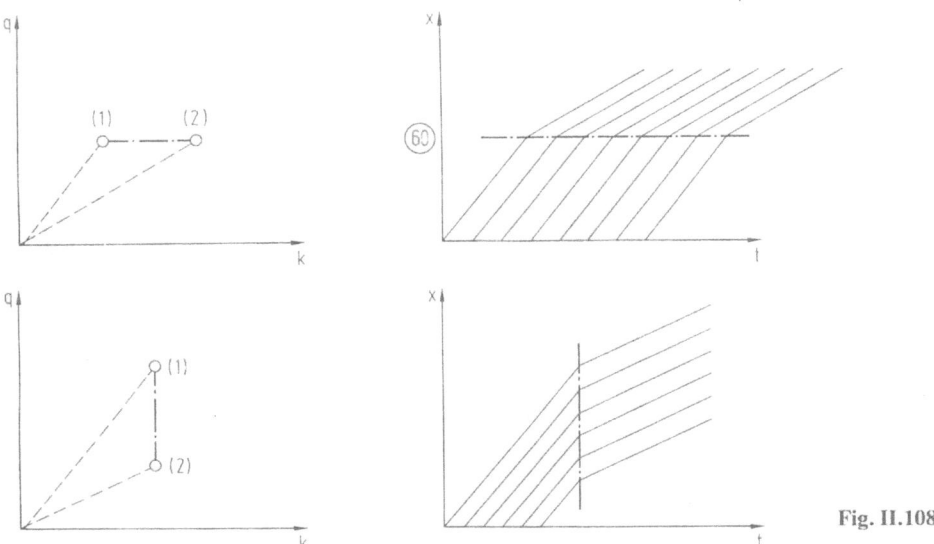

Fig. II.108

Both the transition from free flow to traffic jam as well as the transition from traffic jam to free flow occur along the straight-lined trajectories of the shockwaves. These two shockwave trajectories have different slopes, because the deceleration is in general greater than the subsequent acceleration. Thus the queue is bounded in the x-t plane by the two shockwaves such that a wedge-shaped area results, which continues to spread out until a reduced input flow rate makes the dissipation of the queue possible.

Shockwaves can move up or downstream. The sign determines the direction of movement; if u is positive the shockwave travels downstream, while if u is negative the shockwave travels upstream. Boundary conditions are $u = 0$ and $u \to \infty$.

When $u = 0$, the shockwave stays still (see Fig. II.108). In this case, $q = \text{const}$; the headways between the vehicles remain unchanged despite the change in state. One could imagine such a situation arising if at a speed restriction sign all drivers were to change their speeds instantaneously.

When $u \to \infty$, $k = \text{const}$ (see Fig. II. 108); the distance between the vehicles remain unchanged but there is an instantaneous change in their speeds. Such a situation may for example arise if sleet suddenly causes all vehicles to reduce their speeds simultaneously.

If vehicles come to a halt in a queue, the resulting delay may be quantified as follows:

— As Fig. II.109 shows, the vehicles stop for a certain period. If H is the time stopped for the first vehicle, (q_1, k_1) the traffic flow condition upstream, (q_2, k_2) the condition downstream and k_{max} the traffic density in the stationary queue, then the number of vehicles halting N can be obtained from the geometric properties of triangles as follows:

$$N = \frac{q_1 \cdot q_2 \cdot k_{max} \cdot H}{q_2 (k_{max} - k_1) - q_1 (k_{max} - k_2)} \qquad (II.187)$$

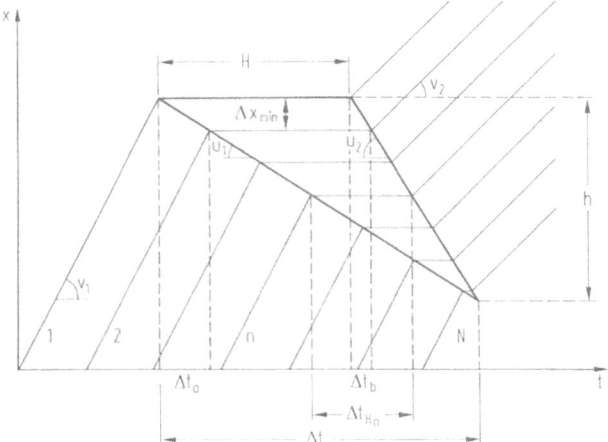

Fig. II.109. (From [100])

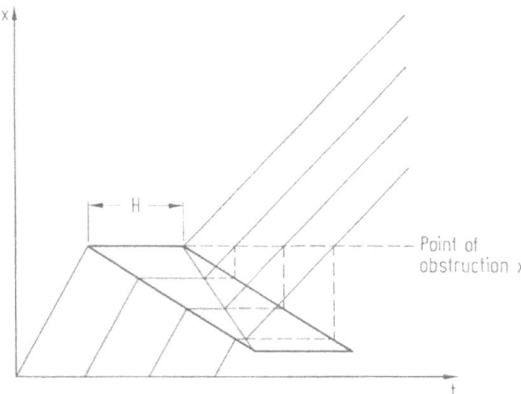

Fig. II.110. (From [100])

(if N is not a whole number, it should be counted down to the next integer).
The total stopped time of all N vehicles is then

$$T_H = N \cdot H - \frac{N(N-1)}{2 \cdot k_{max}} \cdot \left(\frac{k_{max} - k_1}{q_1} - \frac{k_{max} - k_2}{q_2} \right). \qquad (II.188)$$

— Occasionally the time from the halt of a vehicle to when it passes the site of the disturbance is of interest (see Fig. II.110). In the case of traffic lights, for example, this is the stop line. This is known as the waiting time.
The total waiting time for the N vehicles brought to a stop is

$$T_W = N \cdot H - \frac{N(N-1)}{2 \cdot k_{max}} \left(\frac{k_{max} - k_1}{q_1} - \frac{k_{max}}{q_2} \right). \qquad (II.189)$$

— Frequently the difference between the waiting time, mentioned above, and the time the vehicle would have required to cover the same distance in the absence of the disturbance if it maintained its approach speed, is of interest

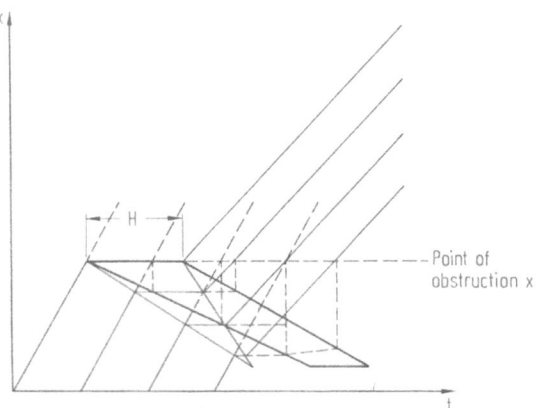

Fig. II.111. (From [100])

(Fig. II.111). This difference is known as lost time. The total lost time for all N vehicles which stop is given by

$$T_v = N \cdot H - \frac{N(N-1)}{2 \cdot k_{max}} \left(\frac{k_{max}}{q_1} - \frac{k_{max}}{q_2} \right). \tag{II.190}$$

II.3.3.3.3 Kinematic Waves or Characteristics

Solutions to Eq. (II.183) exist which are lines of the form

$$x = \int c \, dt + C = ct + x_0$$

along which λ, $\varkappa$ and v are constant. These straight lines are called characteristics or kinematic waves. It is also evident from Eq. (II.185) that the slope c is determined by the tangent to the fundamental diagram $\lambda = \lambda(\varkappa)$ at a point $(\lambda, \varkappa)$, and gives the speed at which kinematic waves are propagated. A kinematic wave can be interpreted as a shockwave resulting from a sufficiently small change in state: as the point $(\lambda_2, \varkappa_2)$ approaches the point $(\lambda_1, \varkappa_1)$ the secant whose slope is $u = \Delta\lambda/\Delta\tau$ approaches the tangent to the fundamental diagram at the point $(\lambda_1, \varkappa_1)$ whose slope is $c = d\lambda/d\varkappa$.

In a time-distance diagram (Fig. II.112b) this is represented by showing the trajectory segments having speed v_2 as approaching the extrapolations of the trajectory segments having speed v_1. From the relationship

$$\tau = \frac{1}{\lambda_i - u\varkappa_i} = \frac{1}{\varkappa_i(v-u)} = \frac{\Delta x_i}{v-u}$$

and since $\lim_{\Delta\varkappa \to 0} u = c$ we obtain

$$\tau' = \frac{\Delta x_1}{v-c} = \frac{1}{\varkappa_1(v-c)} = \frac{1}{\lambda_1 - c\varkappa_1}.$$

Therefore, τ' is the average headway with which vehicles traverse a kinematic wave. Once the limiting process has been carried out, a graphical illustration of

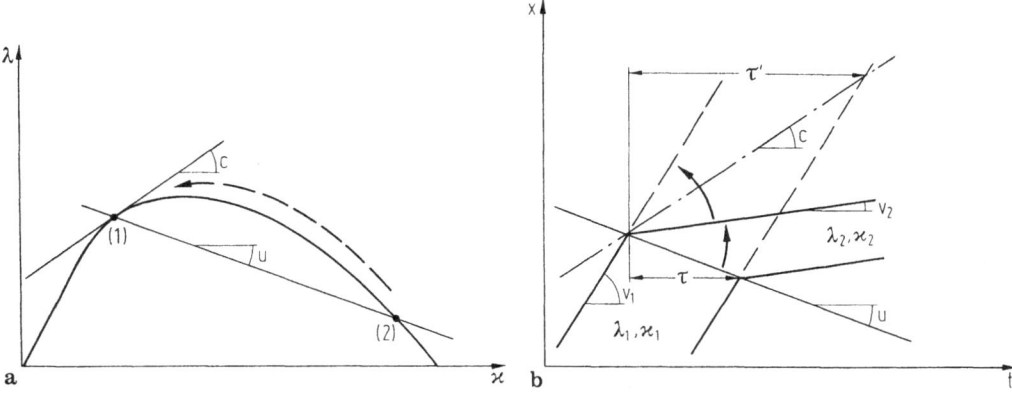

Fig. II.112a, b

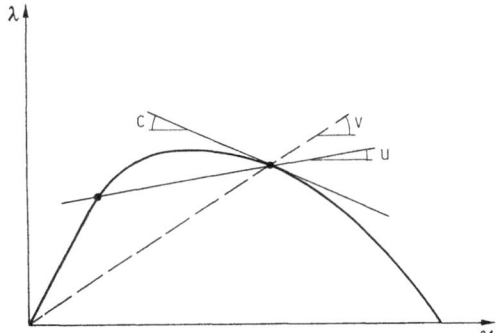

Fig. II.113

this phenomenon is not possible. Figure II.113 summarizes the three speeds treated so far in this discussion in terms of their relationship to the fundamental diagram.

Now let us use the equation of state $\lambda = v\varkappa$ with $v = v(\varkappa)$ (cf. Sect. II.2.5.3.3) in order to derive a new expression for c:

$$c = \frac{d\lambda}{d\varkappa} = v + \varkappa \frac{dv}{d\varkappa}. \tag{II.191}$$

Since $v = v(\varkappa)$ decreases monotonically with increasing density (see Sect. II.2.5.3.2) the slope $dv/d\varkappa \leq 0$ so that $c \leq v$.

The speed of a kinematic wave is therefore always less than the speed of the traffic stream, except in the free flow regime, in which the continuum theory is only a very rough approximation. The tangent to the curve $\lambda(\varkappa)$ shows that a kinematic wave moves with the traffic stream for $v < v_{opt}$ and against the traffic stream for $v > v_{opt}$. At $v = v_{opt}$ the kinematic wave has zero speed, forming a standing wave. For stationary traffic, the kinematic wave trajectories are parallel. But when conditions are not stationary, that is $\varkappa = \varkappa(t)$ with the concentration decreasing over time, a series of kinematic waves results in which waves of smaller concentration (having higher propagation speeds) catch up with waves of higher

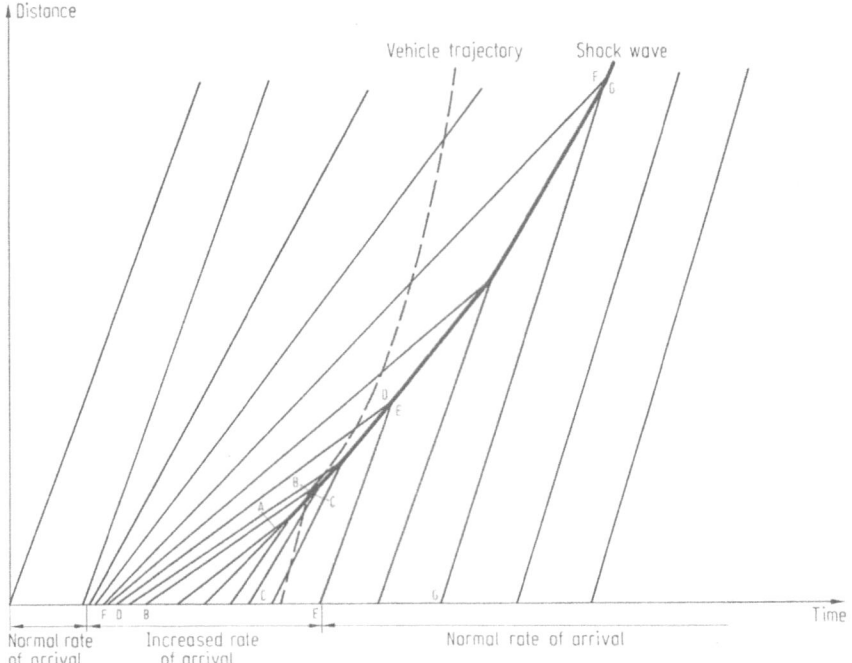

Fig. II.114. (From [70])

concentration (having smaller propagation speeds). At the intersection of two kinematic waves both the concentration as well as the speed undergo a transition. This is referred to as a shockwave, (see Fig. II.114), whose speed is already known to be $u = \Delta\lambda / \Delta\varkappa$.

If the concentration-time function were known at some location x_0 and if the fundamental diagram at this location were also known, then it would be possible to use the continuum theory as demonstrated above to show how the density at x_0, at each point in time, was propagated to other locations. Wherever the resulting straight lines intersect, a shockwave arises whose speed can also be determined. In this fashion, the x-t plane is covered by a series of lines of constant concentration enabling the concentration-time function value for any location to be calculated or ascertained directly (see Figs. II.115 and II.116).

Similar graphical constructions are possible when the concentration-distance function over a section of road is known at a particular time, as would be the case if one had an aerial photograph. However, because of the random characteristics of traffic streams the continuum theory requires, in general, considerable aggregation over time or over distance. A photographic technique which encompasses sufficiently long road sections with sufficiently dense traffic would experience great difficulty. In practice it is very difficult to verify whether traffic flow phenomena are predictable because of the problems of determining the relationship between λ and $\varkappa$, particularly in the regime $v > v_{opt}$.

If $\varkappa = \varkappa(x)$, then c is also a function of concentration and distance: $c = c(\varkappa, x)$.

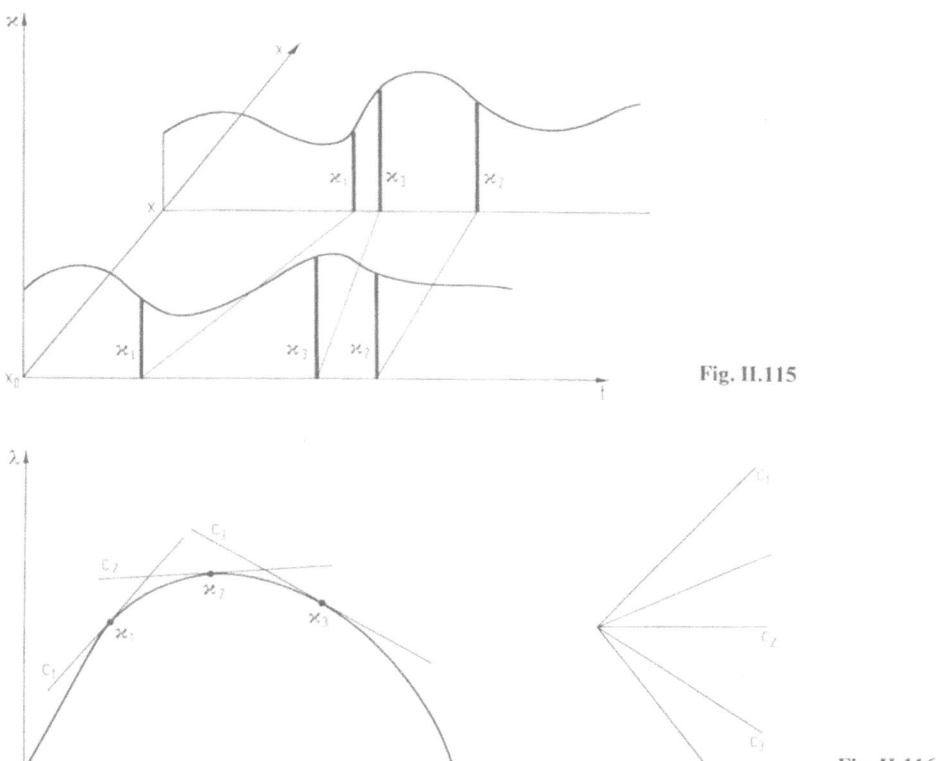

Fig. II.115

Fig. II.116

The trajectories of the kinematic waves are then no longer straight lines but instead curves:

$$\frac{dx}{dt} = c(\varkappa,x) = \frac{\delta\lambda(\varkappa,x)}{\delta\varkappa(x)}$$

$$t = \int_0^x \frac{dx}{c(\varkappa,c)} + C = \int_0^x \frac{\delta\varkappa(x)}{\delta\lambda(\varkappa,x)} dx + C$$

or, when $\varkappa$ is at least piecewise constant, the trajectories are broken lines:

$$t = \sum_{i=1}^n \frac{\Delta x_i}{c(\varkappa,\Delta x_i)} + C.$$

C is the intersection of the kinematic wave with the t-axis.

Example 43.
1. A distance X is composed of two separate sections Δx_1 and Δx_2. For each section there is a different fundamental diagram $\lambda' = \lambda'(\varkappa)$ in Δx_1 and $\lambda'' = \lambda''(\varkappa)$ in Δx_2 (Fig. II.117).

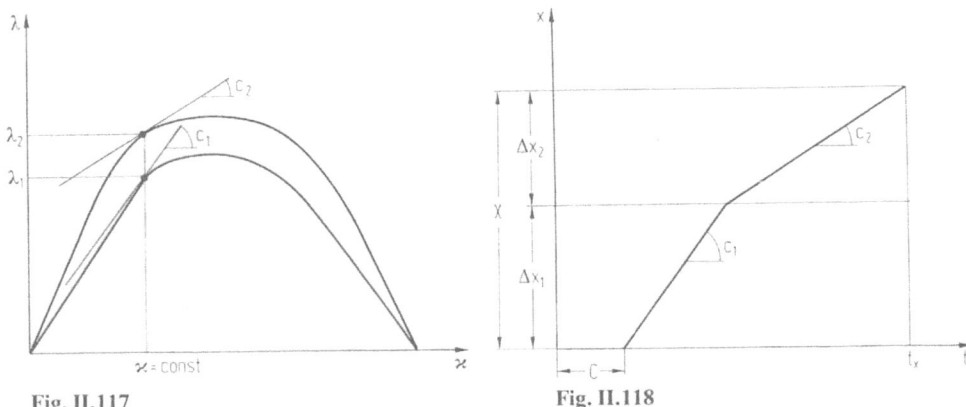

Fig. II.117   Fig. II.118

Let $\varkappa$ be constant. Thus (Fig. II.118)

$$t_x = C + \frac{\Delta x_1}{c_1} + \frac{\Delta x_2}{c_2}.$$

2. Let $\lambda = \lambda(\varkappa,x)$ be given in closed form as a parabola

$$\lambda(\varkappa,x) = -\alpha\varkappa(\varkappa - \varkappa_{max})$$

for which

$$\alpha = 4\lambda_{max}(x)/\varkappa^2_{max},$$

where λ_{max} is a linearly decreasing function of x,

$$\lambda_{max}(x) = \lambda_{max}(x_0) - ax,$$

so that $\alpha = \alpha(x)$ and

$$c(\varkappa,x) = \delta\lambda/\delta\varkappa = -\alpha(x)\ (2\varkappa - \varkappa_{max}).$$

$$\frac{\delta\varkappa}{\delta\lambda} = \frac{1}{\alpha(x)(\varkappa_{max} - 2\varkappa)} = \frac{\varkappa^2_{max}}{4\lambda_{max}(\varkappa_{max} - 2\varkappa)}$$

$$= \frac{\varkappa^2_{max}}{4(\varkappa_{max} - 2)[\lambda_{max}(x_0) - ax]} = \frac{b}{d + ex}$$

where

$$b = \varkappa^2_{max}; \quad d = 4\lambda_{max}(x_0)(\varkappa_{max} - 2\varkappa); \quad e = 8a\varkappa - 4a\varkappa_{max}.$$

Since, so long as $(d + ex) < 0$,

$$\int \frac{dx}{d + ex} = \frac{1}{e}\ln(d + ex) + C$$

then

$$t = \int_0^x \frac{\delta\varkappa}{\delta\lambda}\,dy = \int_0^x \frac{b\,dy}{d + ey} = \frac{b}{e}\ln(d + ey)|_0^x = \frac{b}{e}[\ln(d + ex) - \ln d].$$

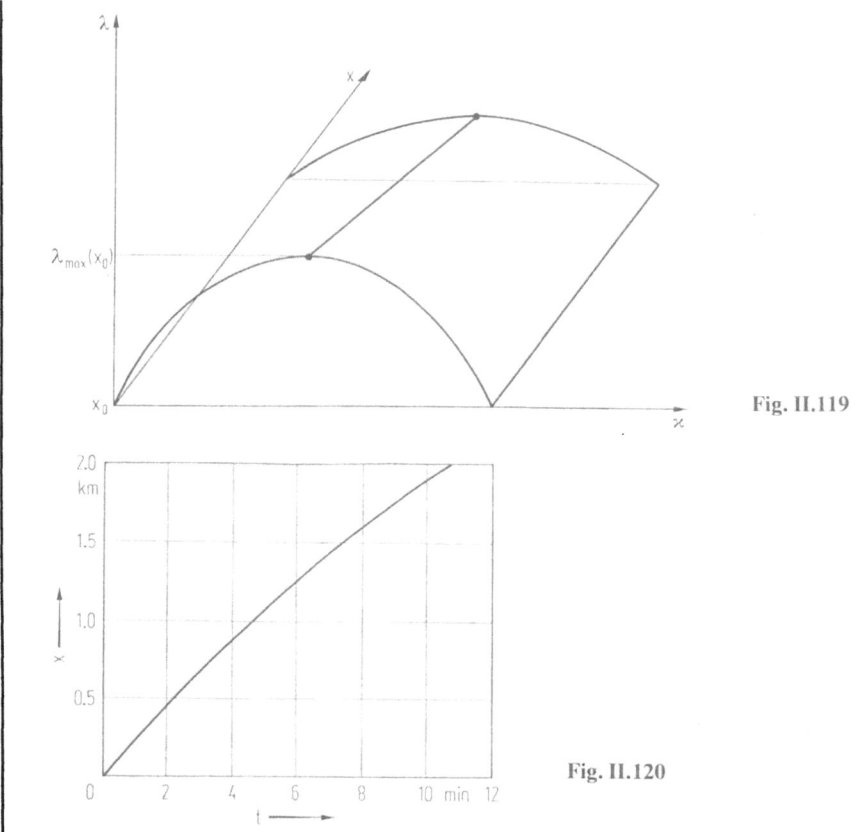

Fig. II.119

Fig. II.120

Figure II.120 shows the behaviour of the kinematic wave for a situation in which $k_{max} = 175$ veh/km, $q_{max}(x_0) = 1500$ veh, $k = 50$ veh/km, and $a = 300$ veh/kmh.

Consider a link on which there is a traffic flow in stationary equilibrium with constant parameters λ, $\varkappa$ and therefore v. This flow is suddenly brought to rest at t_0 either through an accident or a traffic signal installation. Subsequently a queue forms behind this blockage, while in front of it the street is empty. After a certain period, the following density profile arises:

The field of kinematic waves c_i (characteristics) corresponding to this situation is sought.

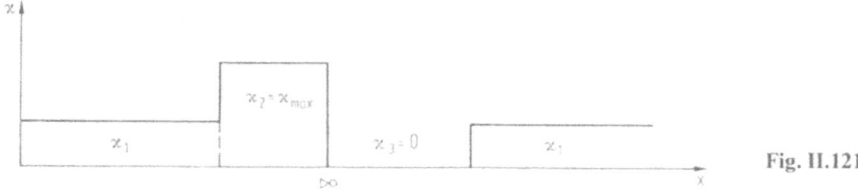

Fig. II.121

For each region $x = x_i$, the respective c_i is obtained from the fundamental diagram. The boundaries between the various traffic flow conditions must be determined. The transition from x_1 to x_2 has already been dealt with in Example 41; the queue grows at a speed corresponding to the shockwave between x_1 and x_{max}.

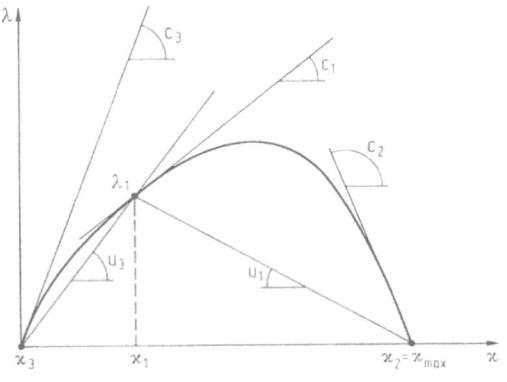

Fig. II.122

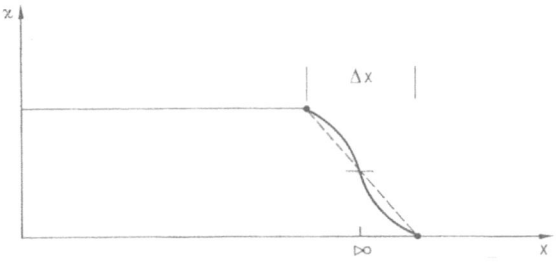

Fig. II.123

The transition from $x_2 = x_{max}$ to $x_3 = 0$ is (because $\lambda_2 = \lambda_3 = 0$) denoted by $u_2 = 0$; the shockwave remains at the site of the interruption to the traffic stream. The transition from $x_3 = 0$ to x_1 is represented by u_3. However, since $x_3 = 0$, u_3 is equal to the radius vector (λ_1, x_1) and therefore equal to v_1:

$$u_3 = \frac{\lambda_3 - \lambda_1}{x_3 - x_1} = \frac{\lambda_1}{x_1} = v_1.$$

Assume now that after a certain time interval τ, namely at $t_1 = t_0 + \tau$, traffic flows freely again (the blockage has been removed or, in the case of traffic signals, the light has changed from red to green). The queue discharges from the front and there is a transition from x_{max} to $x = 0$. Figure II.123 shows this situation at $t_2 = t_1 + \Delta\tau$.

For this transition, the characteristics generated may be obtained from the fundamental diagram.

As $\Delta\tau \to 0$, $\Delta x \to 0$, and a fan-shaped series of characteristics are generated by the transition from x_{max} to $x = 0$. Fan-shaped characteristics cut the characteristics of the incoming traffic both before and after the blockage.

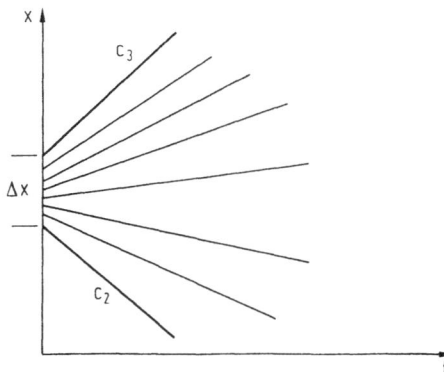

Fig. II.124

In this way, curved shockwaves are formed which run asymptotically into those of the incoming traffic. These may be constructed graphically or calculated analytically from knowledge of the function $v = v(x)$.

Example 44. Given the following linear relationship, which is a crude approximation to reality in traffic flow

$$v(x) = v_{max}[1 - (x/x_{max})]$$

(see Fig. II.125) we have

$$\lambda = x \cdot v = x \cdot v_{max}\left(1 - \frac{x}{x_{max}}\right)$$

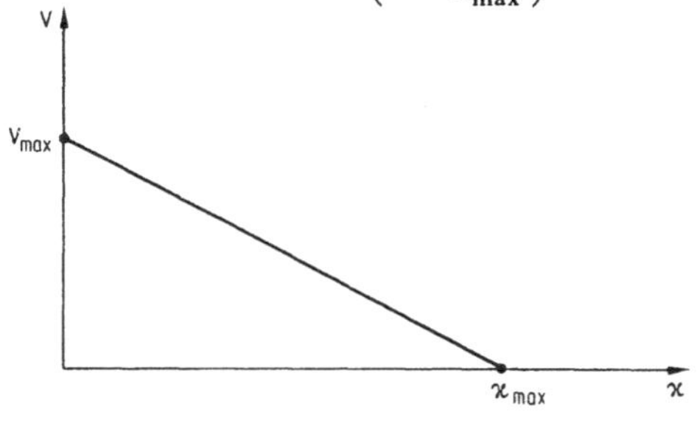

Fig. II.125

and the equation for the characteristic is

$$c = \frac{d\lambda}{dx} = v_{max}\left(1 - \frac{2x}{x_{max}}\right).$$

Denoting the traffic conditions corresponding to the fan-shaped series of characteristics by (λ_f, x_f) and setting $\lambda_f = v \cdot x_f$, we obtain

$$u = \frac{\Delta\lambda}{\Delta x} = \frac{v_{max}\left[x_f\left(1 - \frac{x_f}{x_{max}}\right) - x_1\left(1 - \frac{x_1}{x_{max}}\right)\right]}{x_f - x_1}$$

$$= v_{max}\frac{x_f - x_1 + \frac{x_1^2 - x_f^2}{x_{max}}}{x_f - x_1} = v_{max}\left(1 - \frac{x_1 + x_f}{x_{max}}\right).$$

However, $\varkappa_f$ is a function of x and t.

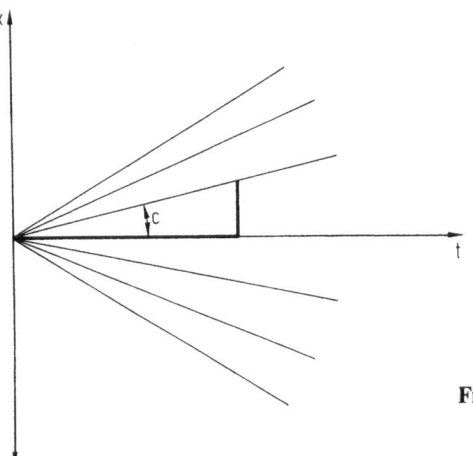

Fig. II.126

Hence

$$c = v_{max} \left(1 - \frac{2\varkappa_f}{\varkappa_{max}} \right) = \frac{x}{t}$$

and thus

$$\varkappa_f = \frac{\varkappa_{max}}{2} \left(1 - \frac{x}{v_{max} \cdot t} \right).$$

If traffic flows freely again at time $t = \tau$, then

$$\varkappa_f = \frac{\varkappa_{max}}{2} \left[1 - \frac{x}{v_{max}(t-\tau)} \right].$$

For non-linear shockwaves we have at any point

$$u = \frac{\Delta\lambda}{\Delta\varkappa} = \frac{dx_s}{dt}$$

where $x_s(t)$ denotes the line of movement of the shockwave. Thus

$$\frac{dx_s}{dt} = v_{max} \left(\frac{1}{2} - \frac{\varkappa_1}{\varkappa_{max}} \right) + \frac{x_s}{2(t-\tau)}.$$

This is an ordinary differential equation with the solution

$$x_s = B(t-\tau)^{1/2} + v_{max} \left(1 - \frac{2\varkappa_1}{\varkappa_{max}} \right)(t-\tau)$$

where B is a constant of integration. This differs between "upper" and "lower" shockwave (see Fig. II.132). The speed of the shockwave is

$$\frac{dx_s}{dt} = v_{max} \left(1 - \frac{2\varkappa_1}{\varkappa_{max}} \right) + \frac{B}{2}(t-\tau)^{-1/2}.$$

For $t \to \infty$, this speed approaches

$$\frac{dx_s}{dt} = v_{max}\left(1 - \frac{2\varkappa_1}{\varkappa_{max}}\right) = c_1$$

namely the speed of the characteristics of the incoming traffic.

The constant of integration B_0 for the "upper" shockwave is determined from the time and distance when the vehicle at the head of the discharging queue travelling at v_{max} meets the undisturbed flow travelling at v_1. Thus

$$v_{max}(t_0 - \tau) = v_1 \cdot t_0$$

and therefore

$$t_0 = \frac{v_{max} \cdot \tau}{v_{max} - v_1}.$$

After insertion of the following assumed linear relationship

$$v_1 = v(\varkappa_1) = v_{max}\left(1 - \frac{\varkappa_1}{\varkappa_{max}}\right)$$

we obtain

$$t_0 = \frac{v_{max} \cdot \tau}{v_{max} - v_{max}\left(1 - \frac{\varkappa_1}{\varkappa_{max}}\right)} = \frac{\tau \cdot \varkappa_{max}}{\varkappa_1}$$

and further

$$x_0 = v_1 \cdot t_0 = v_{max}\left(1 - \frac{\varkappa_1}{\varkappa_{max}}\right) \cdot \frac{\tau \cdot \varkappa_{max}}{\varkappa_1} = v_{max} \cdot \tau \left(\frac{\varkappa_{max}}{\varkappa_1} - 1\right).$$

Substituting this into Eq. (II.192), we obtain

$$B_0 = \frac{\left[x_0 - v_{max}\left(1 - \frac{2\varkappa_1}{\varkappa_{max}}\right)\right](t_0 - \tau)}{(t_0 - \tau)^{1/2}}$$

$$= \frac{v_{max} \cdot \tau \left(\frac{\varkappa_{max}}{\varkappa_1} - 1\right) - v_{max}\left(1 - \frac{2\varkappa_1}{\varkappa_{max}}\right)\left(\frac{\tau \cdot \varkappa_{max}}{\varkappa_1} - \tau\right)}{\left(\frac{\tau \cdot \varkappa_{max}}{\varkappa_1} - \tau\right)^{1/2}}$$

$$= \frac{2 \cdot v_{max} \cdot \tau^{1/2}\left(1 - \frac{\varkappa_1}{\varkappa_{max}}\right)}{\left(\frac{\varkappa_{max}}{\varkappa_1} - 1\right)^{1/2}} = \frac{2 \cdot v_1 \cdot \tau^{1/2}}{\left(\frac{\varkappa_{max}}{\varkappa_1} - 1\right)^{1/2}}.$$

The constant of integration B_u for the lower shockwave is determined by the time and location at which the first vehicle of the in-flowing traffic meets the last vehicle in the queue, just after it recommences to move. In order to determine this point, it is necessary to know how rapidly the queue discharges and how quickly the back of the queue travels backwards.

The speed at which the back of the queue travels backward has already been determined. It is:

$$u_1 = \frac{\lambda_1}{\varkappa_1 - \varkappa_{max}} = \frac{v_1 \cdot k_1}{\varkappa_1 - \varkappa_{max}} = \frac{\varkappa_1 \cdot v_{max}\left(1 - \dfrac{\varkappa_1}{\varkappa_{max}}\right)}{\varkappa_1 - \varkappa_{max}} = -\frac{\varkappa_1 \cdot v_{max}}{\varkappa_{max}}.$$

The speed at which the queue discharges depends on the function $v = v(\varkappa)$. As demonstrated, in the region of the fan-shaped, straight characteristics

$$c = \frac{d\lambda}{d\varkappa} = \frac{x}{t}.$$

On the other hand, for the linear relationship between v and $\varkappa$

$$c = v_{max}\left(1 - \frac{2\varkappa}{\varkappa_{max}}\right).$$

Thus

$$\varkappa = \frac{\varkappa_{max}}{2}\left(1 - \frac{x}{v_{max} \cdot t}\right)$$

yields at each point in time a linear density-distance function:

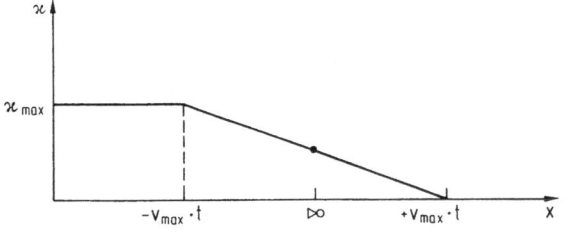

Fig. II.127

$$\text{for } \left\{\begin{array}{l} x = +v_{max} \cdot t \\ x = -v_{max} \cdot t \\ x = 0 \end{array}\right\} \text{ we have } \left\{\begin{array}{l} \varkappa = 0 \\ \varkappa = \varkappa_{max} \\ \varkappa = \dfrac{\varkappa_{max}}{2}. \end{array}\right.$$

No further vehicles come to rest if

$$-\frac{\varkappa_1 \cdot v_{max}}{\varkappa_{max}} \cdot t_u = -v_{max}(t_u - \tau)$$

$$t_u = \frac{\tau}{1 - \dfrac{\varkappa_1}{\varkappa_{max}}}$$

in which case

$$x_u = -v_{max}(t_u - \tau) = -v_{max}\left(\frac{\tau}{1 - \dfrac{\varkappa_1}{\varkappa_{max}}} - \tau\right)$$

$$= -v_{max} \cdot \tau \frac{\varkappa_1}{\varkappa_{max} - \varkappa_1}.$$

When this is in turn substituted into Eq.(II.192), we obtain

$$B_u = \frac{x_u - v_{max}\left(1 - \frac{2x_1}{x_{max}}\right)(t_u - \tau)}{(t_u - \tau)^{1/2}}$$

$$= \frac{-v_{max} \cdot \tau \cdot \frac{x_1}{x_{max} - x_1} - v_{max}\left(1 - \frac{2x_1}{x_{max}}\right)\left(\frac{\tau}{1 - \frac{x_1}{x_{max}}} - \tau\right)}{\left(\frac{\tau}{1 - \frac{x_1}{x_{max}}} - \tau\right)^{1/2}}$$

$$= -2 \cdot v_{max}\left(\frac{x_1 \cdot \tau}{x_{max} - x_1}\right)^{1/2}\left(1 - \frac{x_1}{x_{max}}\right)$$

$$= -2 \cdot v_1 \cdot \tau^{1/2}\left(\frac{x_1}{x_{max} - x_1}\right)^{1/2} = -\frac{2 \cdot v_1 \cdot \tau^{1/2}}{\left(\frac{x_{max}}{x_1} - 1\right)^{1/2}}.$$

The two constants of integration distinguish themselves only by their sign.

When the x-t plane has been filled with characteristics, the movement of an individual vehicle through the plane may be determined.

At each point in the plane

$$dx/dt = v(x,t).$$

Since speed depends on density alone, speed is constant along each characteristic

$$v = v(x).$$

Hence, these lines also correspond to isoquants for the differential equation

$$dx/dt = v(x,t).$$

Areas in the x-t plane in which density is constant correspond to areas where speeds are also constant. However, when density depends on x and t, so does speed.

Example 45. In Example 44 it was assumed that there is a linear relationship between v and x:

$$v = v_{max}\left(1 - \frac{x}{x_{max}}\right).$$

However, in the region of fan-shaped characteristics, x depends on x and t

$$x(x,t) = \frac{x_{max}}{2}\left(1 - \frac{x}{v_{max} \cdot t}\right)$$

(see Example 44). After substitution

$$v(x,t) = \frac{dx}{dt} = v_{max}\left[1 - \frac{1}{2}\left(1 - \frac{x}{v_{max}\cdot t}\right)\right] = \frac{v_{max}}{2} + \frac{x}{2\cdot t}.$$

This is an ordinary non-homogeneous, linear, first order differential equation with the solution

$$x = v_{max}\cdot t + B\cdot t^{1/2}.$$

Figure II.128 shows $v(x,t)$ for a particular numerical example where $v_{max} = 50$ km/h and $\varkappa_{max} = 150$ veh/km.

B is a constant of integration to be obtained from the initial conditions, which related to when the vehicle first sets itself in motion.

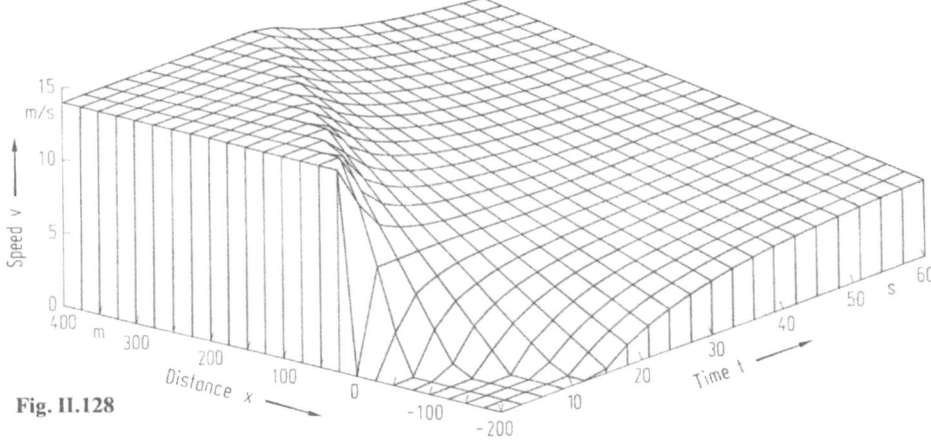

Fig. II.128

This point lies on the characteristic with c_2 (corresponding to $\varkappa_{max}$). In this case $x = -x_0$ and

$$c = \frac{x}{t} = v_{max}\left(1 - \frac{2\varkappa}{\varkappa_{max}}\right)$$

$$c_2 = \frac{x}{t} = v_{max}\left(1 - \frac{2\varkappa_{max}}{\varkappa_{max}}\right) = -v_{max}$$

and therefore

$$t_0 = \frac{-x_0}{-v_{max}} = \frac{x_0}{v_{max}}.$$

This yields

$$-x_0 = x_0 + B\cdot\frac{x_0^{1/2}}{v_{max}^{1/2}}$$

$$B = -2(x_0\cdot v_{max})^{1/2}.$$

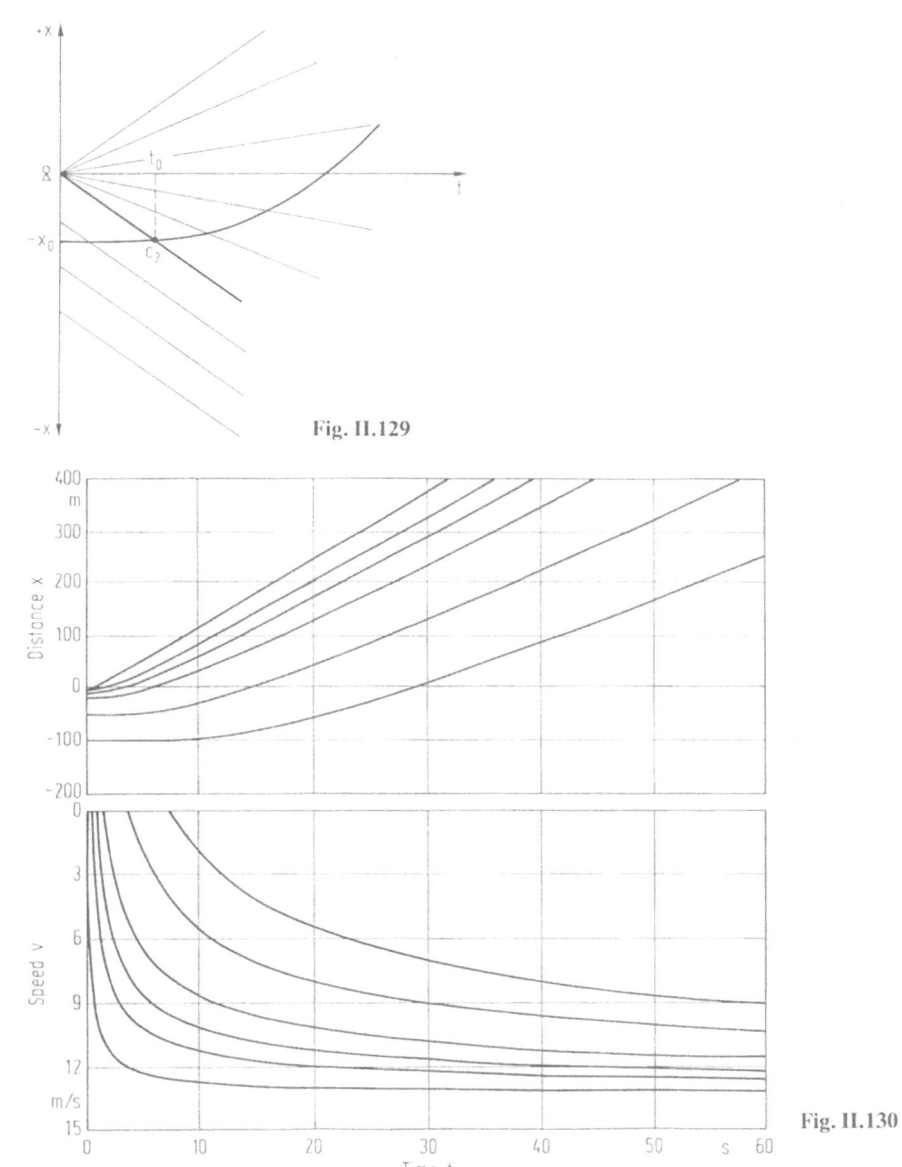

Fig. II.129

Fig. II.130

The equation of motion, or trajectory, of the vehicle under consideration is therefore:

$$x = v_{max} \cdot t - 2 (x_0 \cdot v_{max} \cdot t)^{1/2}$$

and the vehicle's speed at every point is

$$\frac{dx}{dt} = v(x,t) = v_{max} - \left(\frac{x_0 \cdot v_{max}}{t} \right)^{1/2}.$$

Hence speed v_{max} is attained in the limit $\tau \to \infty$.

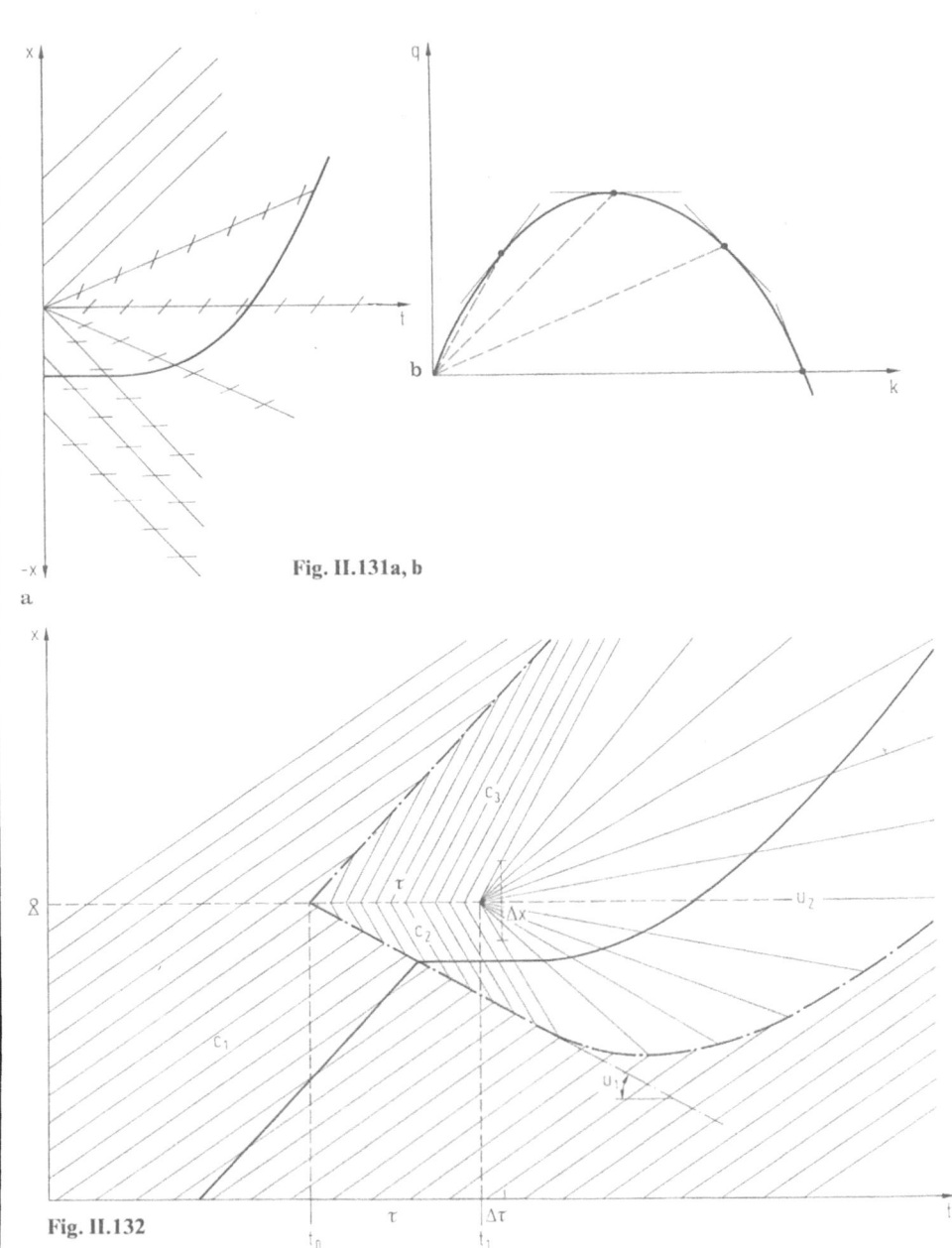

Fig. II.131a, b

Fig. II.132

In Fig. II.130 the time-distance lines and the speed-distance profiles are shown for different starting points $-x_0$.

The trajectory of the vehicle can also be determined graphically from the characteristics.

The starting point is again the fundamental diagram and the field of characteristics (Fig. II.131).

The slope of the characteristics corresponds to the slopes of the tangents of the fundamental diagram, while the speed corresponding to every traffic condition, or point on the fundamental diagram, is characterised by the radius vector through this point (Fig. II.131b). The short, straight lines marked on each characteristic in Fig. II.131a are parallel to the corresponding radius vectors. The equation of motion of a vehicle must, at the point of contact with a characteristic, run parallel to it. Figure II.132 shows the path of a vehicle determined in this way through the field of characteristics resulting from a single interruption to the traffic stream.

II.3.3.3.4 Form of the Fundamental Diagram II

It was shown in Sect. II.3.3.1.4.2 how different combinations of the exponents m and l in the general (microscopic) car following equation [Eq. (II.165)] can lead to different forms of the (mascroscopic) functions $v = v(k)$ and $q = q(k)$. In what follows, the continuity equation will be used to describe different forms of the fundamental diagram.

Let v be a function of x and t:

$$v = v(x,t).$$

The total derivative with respect to time is then

$$\frac{dv}{dt} = \frac{\delta v}{\delta x}\frac{dx}{dt} + \frac{\delta v}{\delta t} = v\frac{\delta v}{\delta x} + \frac{\delta v}{\delta t}.$$

But, since

$$v = v(\varkappa)$$

we have

$$\frac{\delta v}{\delta x} = \frac{dv}{d\varkappa}\frac{\delta \varkappa}{\delta x}$$

and

$$\frac{\delta v}{\delta t} = \frac{dv}{d\varkappa}\frac{\delta \varkappa}{\delta t}.$$

Therefore

$$\frac{dv}{dt} = v\frac{dv}{d\varkappa}\frac{\delta \varkappa}{\delta x} + \frac{dv}{d\varkappa}\frac{\delta \varkappa}{\delta t}.$$

Inserting the equation of continuity

$$\frac{\delta \varkappa}{\delta t} = -c\frac{\delta \varkappa}{\delta x}$$

we obtain

$$\frac{dv}{dt} = (v-c)\frac{dv}{d\varkappa}\frac{\delta \varkappa}{\delta x}.$$

With Eq. (II.191)

$$c = v + \varkappa \frac{dv}{d\varkappa}$$

the final result is

$$\frac{dv}{dt} = \left(v - v - \varkappa \frac{dv}{d\varkappa} \right) \frac{dv}{d\varkappa} \frac{\delta\varkappa}{\delta x} = - \varkappa \left(\frac{dv}{d\varkappa} \right)^2 \frac{\delta\varkappa}{\delta x}$$

or, with

$$- \varkappa \left(\frac{dv}{d\varkappa} \right)^2 = F$$

(II.192)

$$\frac{dv}{dt} = F \frac{\delta\varkappa}{\delta x}.$$

The last equation describes the acceleration of vehicles in a traffic stream as a function of the derivative of the concentration with respect to distance. Because $dv/d\varkappa$ is squared, the sign of the acceleration does not depend on $v = v(\varkappa)$ but rather on $\delta\varkappa/\delta x$:

if $\dfrac{\delta\varkappa}{\delta x} < 0$, the traffic stream is moving from a region of higher concentration into a region of lower concentration, so that the acceleration is positive;

if $\dfrac{\delta\varkappa}{\delta x} > 0$, the traffic stream is moving from a region of lower concentration into a region of higher concentration, so that the acceleration is negative (the drivers must brake);

if $\dfrac{\delta\varkappa}{\delta x} = 0$, the concentration is stationary over distance, so that the acceleration is equal to zero and therefore the speed of the traffic stream is stationary (see Sect. II.3.1.1).

For $\varkappa = \varkappa(x)$, therefore, not only the trajectories of the density waves, but also the trajectories of the vehicles are curves instead of straight lines.

Since the value of F depends upon $dv/\varkappa$ each $v(\varkappa)$ — and therefore each value of $\lambda(\varkappa) = \varkappa v(\varkappa)$ — corresponds to a distinct value of F. Two examples can be used for illustration:

1. If

$$v = v_{opt} \ln \frac{\varkappa_{max}}{\varkappa}$$

[see Eq. (II.169)], then

$$\frac{dv}{d\varkappa} = \frac{v_{opt}}{\varkappa}$$

and thus

$$F = - \frac{v_{opt}^2}{\varkappa}.$$

Therefore,

$$\frac{dv}{dt} = - \frac{v_{opt}^2}{\varkappa} \frac{\delta \varkappa}{\delta x}.$$

But this is the equation of motion of a one-dimensional fluid with the state parameter v_{opt}. This case corresponds to the general model of car-following with $m=0$, $l=1$ (see Sect. II.3.3.1.4).

2. If

$$v = v_w \left(1 - \frac{\varkappa}{\varkappa_{max}} \right)$$

[see Eq. (II.171)], then

$$\frac{dv}{d\varkappa} = - \frac{v_w}{\varkappa_{max}}$$

and thus

$$F = - \varkappa \frac{v_w^2}{\varkappa_{max}^2}.$$

This corresponds to the general car-following model with $m=0$, $l=2$ (see Sect. II.3.3.1.4).

If, instead of the total derivative, we take the partial derivative of speed with respect to time

$$\frac{\delta v}{\delta t} = \frac{dv}{d\varkappa} \frac{\delta \varkappa}{\delta t}$$

and substitute

$$\frac{\delta \varkappa}{\delta t} = - c \frac{\delta \varkappa}{\delta x}$$

one obtains

$$\frac{\delta v}{\delta t} = - c \frac{dv}{d\varkappa} \frac{\delta \varkappa}{\delta x}.$$

We then have the acceleration of a traffic stream (or of one vehicle in the stream) as it would be observed at a fixed location. The acceleration depends upon $v(\varkappa)$. If $dv/d\varkappa = 0$, then the acceleration is equal to zero. This holds at the point $v = v_w$.

Since, elsewhere, $dv/d\varkappa$ is negative, the signs of c and $\delta \varkappa / \delta x$ determine the sign of the acceleration. It can be negative or positive, except when $c=0$ at $v = v_{opt}$, at which point the locally observed speed is also equal to zero.

Setting

$$-c\frac{dv}{d\varkappa}=G,$$

then

$$G=-\left(v+\varkappa\frac{dv}{d\varkappa}\right)\frac{dv}{d\varkappa}=-v\frac{dv}{d\varkappa}-\varkappa\left(\frac{dv}{d\varkappa}\right)^2=F-v\frac{dv}{d\varkappa}$$

so that

$$\frac{\delta v}{\delta t}=G\frac{\delta\varkappa}{\delta x}. \tag{II.193}$$

II.3.3.3.5 The Application of Continuum Theory to a Multi-lane Carriageway

The continuum model may be generalised to handle the distribution of traffic between lanes by specifying the continuity equations for each lane and by permitting certain density fluctuations across lanes. The "density exchange" between neighbouring lanes is determined by the density in the two lanes.

Consider an n-lane carriageway. Let $\varkappa_i(x,t)$ represent the density function in lane i and $\varkappa_{i0}$ the equilibrium densities in lane i at which no further density fluctuations occur. Further, let

$$K_i(x,t)=\varkappa_i(x,t)-\varkappa_{i0}$$

be the derivative of the density from this equilibrium value ("density disturbance" in lane i). When it is assumed that under equilibrium the same density occurs in every lane, a continuity equation for the density disturbance in every lane is obtained [see Eqs.(II.184) and (II.185)]:

$$\frac{\delta K_1}{\delta t}+c\frac{\delta K_1}{\delta x}=a(\varkappa_2-\varkappa_1)$$

$$\frac{\delta K_i}{\delta t}+c\frac{\delta K_i}{\delta x}=a(\varkappa_{i-1}-2\varkappa_i+\varkappa_{i+1})\qquad(i=2,3,\ldots,n-1)\qquad(II.194)$$

$$\frac{\delta K_n}{\delta t}+c\frac{\delta K_n}{\delta x}=a(\varkappa_{n-1}-\varkappa_n)$$

where a is a positive constant with the dimension time^{-1}.

For the system of Eq.(II.194), matrix notation may be introduced:

$$(D_t+cD_x)K+aAK=0 \tag{II.195}$$

where D_t and D_x are operators for partial differentiation with respect to time and distance respectively, K is a vector of density disturbances:

$$K^T=(K_1,\ldots,K_n)$$

and A is a symmetric matrix

$$
A = \begin{pmatrix}
1 & -1 & 0 & 0 & \cdots\cdots & 0 \\
-1 & 2 & -1 & 0 & \cdots\cdots & 0 \\
0 & -1 & 2 & -1 & \cdots\cdots & 0 \\
\vdots & & & & & \vdots \\
0 & \cdots\cdots & & -1 & 2 & -1 \\
0 & \cdots\cdots & & 0 & -1 & 1
\end{pmatrix}.
$$

Let us consider the following fundamental relationship:

$$
\lambda = \lambda(\varkappa) = \begin{cases}
\dfrac{\lambda_0}{\varkappa_0}\,\varkappa, & \varkappa \leqq \varkappa_0 \\[2ex]
\dfrac{\lambda_0}{1 - \dfrac{\varkappa_0}{\varkappa_{max}}}\left(\dfrac{\varkappa_{max}}{\varkappa} - 1\right), & \varkappa > \varkappa_0
\end{cases}
\tag{II.196}
$$

$(\varkappa_0, \lambda_0)$ identifies the point in the fundamental diagram at which maximum traffic flow arises, and $\varkappa_{max}$ identifies maximum density. The fundamental relationship defined in Eq. (II.196) is portrayed in Fig. II.133.

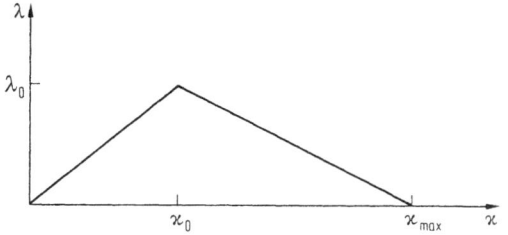

Fig. II.133

The solution to Eq. (II.195) is

$$
K(x,t) = M \cdot B(x) M^{-1} K\left(0, t - \frac{x}{c}\right)
\tag{II.197}
$$

where M is an orthogonal matrix with

$$
M \cdot A \cdot M^{-1} = S = \begin{pmatrix}
\lambda_1 & 0 & \cdots\cdots & 0 \\
0 & \lambda_2 & & \vdots \\
\vdots & \vdots & \ddots & \vdots \\
0 & 0 & \cdots\cdots & \lambda_n
\end{pmatrix}.
$$

and $\lambda_1, \ldots, \lambda_n$ are the eigen values of matrix A.

Defining $d_i(x) = \exp(-\lambda_i ax/c)$ vector B has the form

$$
B^T(x) = (d_1(x), \ldots, d_n(x)).
$$

For example, when $n = 4$:

$$M = \begin{pmatrix} 1 & 1 & 1 & 1 \\ 1 & \sqrt{2}-1 & -1 & -\sqrt{2}-1 \\ 1 & 1-\sqrt{2} & -1 & \sqrt{2}+1 \\ 1 & -1 & 1 & -1 \end{pmatrix}$$

$$M^{-1} = \frac{1}{8} \begin{pmatrix} 2 & 2 & 2 & 2 \\ 2+\sqrt{2} & \sqrt{2} & -\sqrt{2} & -2-\sqrt{2} \\ 2 & -2 & -2 & 2 \\ 2-\sqrt{2} & \sqrt{2} & \sqrt{2} & 2+\sqrt{2} \end{pmatrix}$$

and

$$\lambda_1 = 0, \quad \lambda_2 = 2-\sqrt{2}, \quad \lambda_3 = 2, \quad \lambda_4 = 2+\sqrt{2}.$$

Taking

$$\begin{pmatrix} \alpha(t) \\ \beta(t) \\ \gamma(t) \\ \delta(t) \end{pmatrix} = M^{-1} \cdot \begin{pmatrix} K_1\left(0, t-\frac{x}{c}\right) \\ K_2\left(0, t-\frac{x}{c}\right) \\ K_3\left(0, t-\frac{x}{c}\right) \\ K_4\left(0, t-\frac{x}{c}\right) \end{pmatrix}$$

and

$$u_1 = \sqrt{2}-1, \quad u_2 = \sqrt{2}+1$$

we obtain the following from Eq. (II.197):

$$K(x, t) = \begin{pmatrix} d_1\alpha + d_2\beta + d_3\gamma + d_4\delta \\ d_1\alpha + u_1 d_2\beta - d_3\gamma - u_2 d_4\delta \\ d_1\alpha - u_1 d_2\beta - d_3\gamma + u_2 d_4\delta \\ d_1\alpha - d_2\beta + d_3\gamma - d_4\delta \end{pmatrix}. \tag{II.198}$$

If, for example, the initial distribution of density is $(\varkappa_i(0,t))$ and hence the vector $K(0,t)$ is known, then the behaviour of density as a function of distance and time may be determined from Eq. (II.198).

II.3.3.3.6 Generalization of the Continuum Theory

Although limited to dense traffic, the continuum theory in the form so far discussed includes some assumptions which obviously do not agree with reality.

1. The theory assumes that changes in speed occur instantaneously upon the passage of a shockwave (see Fig. II.79); it neglects reaction times, in the broadest sense, as well as the time necessary for deceleration and acceleration.
2. The theory assumes that reactions to changes in concentration will occur only after the region of changed concentration is entered; it therefore neglects the fact that drivers perceive such changes in advance and are able to react in a precautionary fashion.
3. Instabilities cannot be explained by the theory.

A first extension of the theory is to assume that the intensity depends not only upon the concentration but also upon its spatial derivative:

$$\lambda = \lambda \left(\varkappa, \frac{\delta \varkappa}{\delta x} \right).$$

Then

$$\frac{\delta \varkappa}{\delta x} = \frac{\delta \lambda}{\delta \varkappa} \frac{\delta \varkappa}{\delta x} + \frac{\delta \lambda}{\delta \left(\dfrac{\delta \varkappa}{\delta x} \right)} \frac{\delta \left(\dfrac{\delta \varkappa}{\delta x} \right)}{\delta x} = c \frac{\delta \varkappa}{\delta x} + \mu \frac{\delta^2 \varkappa}{\delta x^2}.$$

Inserting this formula into the equation of continuity we obtain

$$\frac{\delta \varkappa}{\delta t} + c \frac{\delta \varkappa}{\delta x} + \mu \frac{\delta^2 \varkappa}{\delta x^2} = 0$$

(II.199)

$$\frac{\delta \varkappa}{\delta t} + c \frac{\delta \varkappa}{\delta x} = -\mu \frac{\delta^2 \varkappa}{\delta x^2}.$$

This is one type of a diffusion equation; μ is called the diffusion coefficient. It accounts for the fact that concentration changes are previewed by drivers. The disturbance term disappears when $\mu = 0$ and the diffusion equation becomes once again the original continuity equation.

The stability characteristics of this equation can be investigated by the following method. Let

$$\varkappa(x,t) = \varkappa_0 e^{i\beta(x - ct)}$$

(II.200)

with c = complex wave speed; β = (real) wave number = $2\pi/L$ (L = wave length). The wave speed c consists of two parts:

$$c = c_p + ic_t.$$

(II.201)

c_p is the physical speed of the wave (= phase speed) and c_t is the measure of the change of amplitude per unit time. Equation (II.200) describes an undulating process of the concentration with respect to time and distance.

Harmonic input functions are frequently used to investigate stability. Since

$$\varkappa(x,t) = \varkappa_0 [\cos \beta(x - ct) + i \sin \beta(x - ct)] = \varkappa_0 e^{i\beta(x - ct)}$$

the exponential form will be used instead of the trigonometrical form; only the real term need be considered.

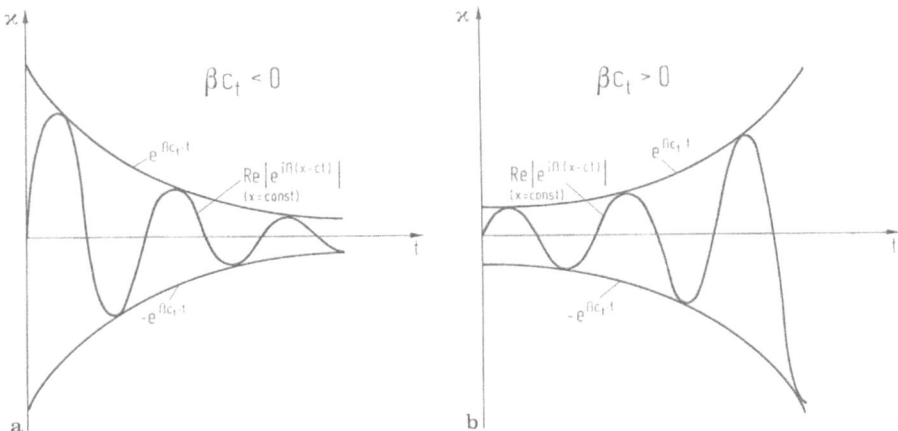

Fig. II.134a, b

Inserting Eq. (II.201) into Eq. (II.200), we obtain

$$x(x,t) = x_0 e^{i\beta[x-(c_p+ic_t)t]} = x_0 e^{i\beta(x-c_p t)} e^{\beta c_t t}. \tag{II.202}$$

Since it is possible to separate the exponent into real and imaginary parts, Eq. (II.202) allows the investigation of the stability behaviour of Eq. (II.199).

Stability exists for the condition (Fig. II.134a)

$$\beta c_t \leqq 0.$$

Instability exists for the condition (Fig. II.134b)

$$\beta c_t > 0.$$

For

$$\beta c_t = 0$$

the amplitude of the oscillation remains constant. Since β can take on only positive values, stability behaviour is therefore controlled by the sign of c_t. From

$$x(x,t) = x_0 e^{i\beta(x-c_p t)} e^{\beta c_t t}$$

it follows that

$$\frac{\delta x}{\delta t} = (\beta c_t - i\beta c_p)(e^{(\beta c_t - i\beta c_p)t + i\beta x}),$$

$$\frac{\delta x}{\delta x} = i\beta e^{(\beta c_t - i\beta c_p)t + i\beta x},$$

$$\frac{\delta^2 x}{\delta x^2} = -\beta^2 e^{(\beta c_t - i\beta c_p)t + i\beta x}.$$

Inserting these derivatives into Eq. (II.199) and cancelling out the common exponential factor results in

$$\beta c_t - i\beta c_p + i\beta c - \mu\beta^2 = 0$$

or

$$-i\beta(c_p - c) + \beta c_t - \mu\beta^2 = 0. \tag{II.203}$$

Let us first set μ equal to zero, hereby obtaining

$$-i\beta(c_p - c) + \beta c_t = 0.$$

This expression can be equal to zero only if both the imaginary and the real parts are separately equal to zero. In that case, $c_p = c$ and $c_t = 0$. Not only does Eq. (II.199), with $\mu = 0$, give us the original continuity equation, as mentioned before, but also, with a solution of the form of Eq. (II.200), it describes the oscillation of a stable undamped wave since, for $\mu = 0$, $c_t = 0$. Such a wave is called a kinematic wave. Let us now consider $\mu \neq 0$. Once again, Eq. (II.203) can be equal to zero only if the real and imaginary parts are both equal to zero. From the resulting condition that $c_p - c = 0$, it follows that the phase speed c_p of the wave corresponds to the wave speed c in Eq. (II.199).

For the imaginary part, the condition that $\beta \cdot c_t - \mu\beta^2 = 0$ leads to the result that $c_t = \mu\beta$.

Thus, if $\mu < 0$, then c_t is negative, so that the kinematic wave is damped and stable.

If $\mu > 0$, then c_t is positive, so that the wave is oscillating with ever-increasing amplitude, which means that it is unstable.

If we assume, in addition, that the intensity depends not only on the concentration and on its derivative with respect to distance, but also upon its derivative with respect to time

$$\lambda = \lambda\left(\varkappa, \frac{\delta\varkappa}{\delta t}, \frac{\delta\varkappa}{\delta x}\right),$$

then we can write

$$\frac{\delta\lambda}{\delta x} = \frac{\delta\lambda}{\delta\varkappa}\frac{\delta\varkappa}{\delta x} + \frac{\delta\lambda}{\delta\left(\frac{\delta\varkappa}{\delta x}\right)}\frac{\delta\left(\frac{\delta\varkappa}{\delta x}\right)}{\delta x} + \frac{\delta\lambda}{\delta\left(\frac{\delta\varkappa}{\delta t}\right)}\frac{\delta\left(\frac{\delta\varkappa}{\delta t}\right)}{\delta x}$$

$$= c\frac{\delta\varkappa}{\delta x} + \frac{\delta^2\varkappa}{\delta x^2} + \frac{\delta^2\varkappa}{\delta t\delta x}.$$

Inserting this result into the equation of continuity, we obtain

$$\frac{\delta\varkappa}{\delta t} + c\frac{\delta\varkappa}{\delta x} + \mu\frac{\delta^2\varkappa}{\delta x^2} + v\frac{\delta^2\varkappa}{\delta t\delta x} = 0. \tag{II.204}$$

By means of v it is possible to model a reaction time delay to changes in concentration.

Once again letting

$$\varkappa(x,t) = e^{i\beta(x - c_p t)}e^{\beta c_t t}$$

we can obtain

$$\frac{\delta^2\varkappa}{\delta t\delta x} = (\beta c_t - i\beta c_p)i\beta e^{(\beta c_t - i\beta c_p)t + i\beta x}.$$

With this result and with the previously derived expressions for the partial derivatives of $\varkappa$, Eq. (II.204) leads to the condition

$$\beta c_t - i\beta c_p + i\beta c - \mu\beta^2 + iv\beta(\beta c_t - i\beta c_p) = 0$$

from which we find that

$$\beta c_t - i\beta c_p = \frac{\beta^2(\mu - cv) - i\beta(c + v\mu\beta^2)}{1 + v\beta^2}.$$

Thus, the sign of the term βc_t depends upon the expression $(\mu - cv)$. If the amplitude of a wave is to decrease with time, then $\beta c_t < 0$ so that it is necessary that $c > \mu/v$. If the amplitude of a wave is to increase with time, then $\beta c_t > 0$, so that it is necessary that $c < \mu/v$. For $c = \mu/v$, the amplitude of the wave remains constant; this is again the case of a kinematic wave, as described by the original equation of continuity.

The generalizations of the original equation of continuity are particularly useful in order to explain why in Sect. II.3.3.3.3 the straight lines of constant density were called kinematic waves. At the moment it must be left undecided, whether, in addition, it is possible to describe quantitatively and in a meaningful way the observed instabilities in a traffic stream by means of the generalised equation of continuity.

II.3.3.3.7 A Dynamic Continuum Model

In contrast to the classic continuum model, in which the continuity equation

$$\frac{\delta\varkappa}{\delta t} + \frac{\delta\lambda}{\delta x} = 0 \tag{II.205}$$

is complemented by a static fundamental relationship $\lambda = \lambda(\varkappa)$, in the dynamic continuum model the continuity equation is complemented by an equation to describe acceleration behaviour:

$$\frac{dv}{dt} = \gamma(V(\varkappa) - v) - \frac{c_0^2}{\varkappa} \cdot \frac{\delta\varkappa}{\delta x}. \tag{II.206}$$

Equation (II.206) contains the relaxation term $\gamma[V(\varkappa) - v]$ and the anticipation term $-(c_0^2/\varkappa)\cdot(\delta\varkappa/\delta x)$. The relaxation term allows for the delayed adjustment of the stream to a prespecified speed $V(\varkappa)$ as a result of reaction time and braking or acceleration procedures. The anticipation term allows for the fact that drivers adjust their speeds in advance to changes in density lying ahead. The coefficient c_0 is a constant of proportionality which corresponds to the speed at which disturbances propagate themselves upstream when traffic density is very high. The relaxation and anticipation effects were already considered in the generalized continuum model already formulated in Sect. II.3.3.3.5.

The continuum model specified in Eqs. (II.205) and (II.206) is a system of differential equations for the functions $\varkappa(x,t)$ and $v(x,t)$. If the initial distribution and the time series is known for a particular point, the time series may be determined for any point. The system of equations can be treated by the methods of theoretical physics. Solutions to the systems of equations (II.205) and (II.206) are

$$\varkappa = \varkappa_0, \quad v = V(\varkappa_0).$$

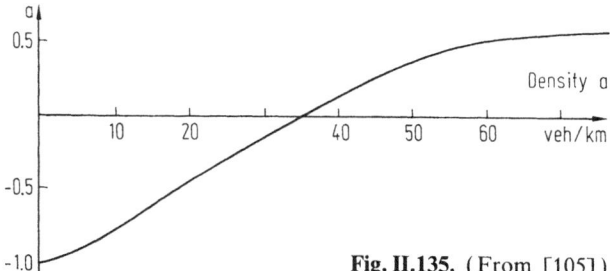

Fig. II.135. (From [105])

In order to investigate the stability of these solutions, one considers

$$\varkappa = \varkappa_0 + \tilde{\varkappa}e^{ikx' + \omega t'}; \quad v = V(\varkappa_0) + \tilde{v}e^{ikx' + \omega t'}$$

where

$$x' = \frac{x \cdot c_0}{\gamma}; \quad t' = \gamma \cdot t.$$

An examination of stability shows that for $a > 0$ the solutions are always instable, independent of the wave number k. For $a < 0$, the solutions are always stable

$$a = -1 - \frac{\varkappa_0}{c_0} \cdot \frac{dV(\varkappa)}{\delta\varkappa}.$$

Parameter a may be represented as a function of traffic density $\varkappa$. For the relationship used in this example

$$V(\varkappa) = \left(v_w \left(1 - \frac{\varkappa}{\varkappa_{max}} \right)^{n_1} \right)^{n_2}$$

with $v_w = 120$ km/h, $\varkappa_{max} = 200$ veh/km/lane, $n_1 = 1.4$, $n_2 = 4$ we obtain the relationship portrayed in Fig. II.135. Note that the critical traffic density between stable and unstable traffic flow in this example is 35 veh/lane/km.

In order to determine the solution of the partial differential equations (II.205) and (II.206) in the region of instability ($a > 0$), one transforms the coordinates

$$x' = x'(x,t); \quad t' = t$$

with

$$\frac{\delta x'}{\delta t} = -\varkappa \cdot v, \quad \frac{\delta x'}{\delta x} = \varkappa$$

and

$$\delta_t = \delta_{t'} - \varkappa v \delta_x, \quad \delta_x = \varkappa \delta_x.$$

After introducing an inverse density function $\chi = 1/\varkappa$, the dynamic continuum model may be written as follows

$$\chi_t - v_x = 0 \tag{II.207}$$

$$v_t + \left(\frac{1}{\chi} \right) x = V\left(\frac{1}{\chi} \right) - v. \tag{II.208}$$

Variables x' and t' are replaced by x and t.

To solve this system of equations, set

$$\chi(x,t) = \chi(y); \quad v(x,t) = v(y)$$

where $y = x + Q \cdot t$. This implies that one is looking for solutions in which the density profile propagates itself with constant speed Q. This leads to the following ordinary system of differential equations

$$(Q \cdot \chi - v)_y = 0 \tag{II.209}$$

$$Q v_y + \left(\frac{1}{\chi}\right)_y = V\left(\frac{1}{\chi}\right) - v. \tag{II.210}$$

Equation (II.209) may be integrated directly

$$v = Q \cdot \chi + v_g \tag{II.211}$$

where v_g is a constant of integration. Substituting $1/\varkappa$ for χ yields

$$\varkappa(v - v_g) = Q. \tag{II.212}$$

Equation (II.212) indicates that the traffic flow Q is stationary in a system moving at speed v_g, which in turn suggests that density increases at the same rate as speed falls relative to v_g.

From Eqs. (II.210) and (II.211) we obtain

$$\left(Q^2 - \frac{1}{\chi^2}\right)\chi_y = V\left(\frac{1}{\chi}\right) - Q\chi - v_g. \tag{II.213}$$

As the factor $Q^2 - (1/\chi^2)$ vanishes so must also the right hand side of Eq. (II.213). This factor disappears if $\chi = 1/Q$, in which case

$$v_g = V(Q) - 1. \tag{II.214}$$

Substituting Eq. (II.214) into Eq. (II.213), Eq. (II.213) may be integrated. The results for different values of Q are shown in Fig. II.136. The figure shows the behaviour of inverse density in a coordinate system moving at v_g. Note that inverse density increases with time. This suggests that in relation to the fixed coordinate system traffic density decreases with time.

This in turn demonstrates that starting from the homogeneous solution $\varkappa = \varkappa_0$, $v = V(\varkappa_0)$ there is no periodic solution for $a > 0$.

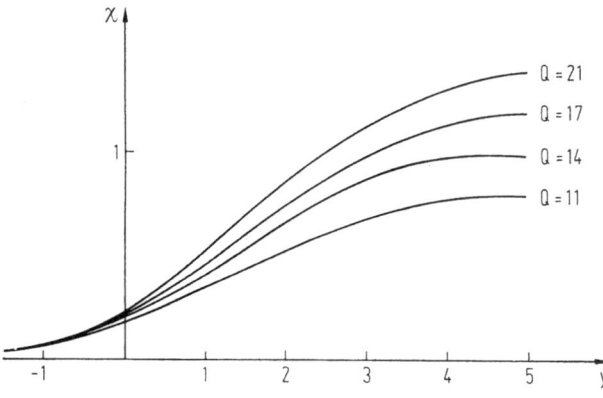

Fig. II.136. (From [105])

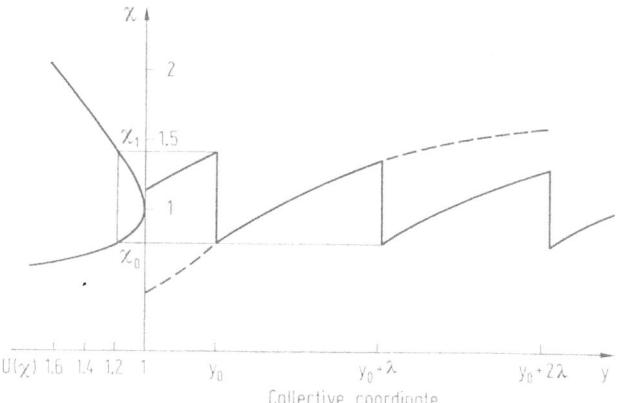

Fig. II.137. (From [105])

Periodic solutions would be possible if finite jumps in traffic density are permitted. This is by no means impossible since, because of the discrete nature of vehicles in a time-distance plane, discrete jumps in the continuity equation are likely to be the rule. Let χ_0 and χ_1 be the inverse densities between which a discrete jump arises. By integrating at the point of the jump over an infinitely small interval, we obtain a jump condition:

$$\chi_0 + \frac{1}{\chi_0} = \chi_1 + \frac{1}{\chi_1}.$$

Hence it is possible to construct continuous solutions step by step. The beginning and end points of each solution step always lie on the line

$$U(\chi) = \chi + \frac{1}{\chi}.$$

A stepwise solution is shown in Fig. II.137.

II.3.3.3.8 Derivation of the Continuum Model from a Kinetic Model

While continuum models describe traffic flow best at higher traffic densities, kinetic models perform best at lower or medium traffic densities. An approach to the formulation of a kinetic traffic flow model that also yields good results in the region of higher traffic densities is to include headway behaviour and vehicle lengths in the kinetic equations.

In this context, the speed-dependent "effective" distance is relevant

$$l = l_0 \left(1 + \frac{v}{c} \right)$$

where l_0 is "effective" length at zero speed (vehicle length and safety distance in a stationary queue) (for comparison see here Sect. II.3.3.2). If $\varkappa_{max}$ is the maximum vehicle density per lane and n is the number of lanes, then

$$l_0 = n/\varkappa_{max}.$$

c is a constant of proportionality.

The Boltzmann equation which results from this equation is:

$$\frac{\delta\varkappa(x,t)f(x,t,v)}{\delta t} + v(x,t)\frac{\delta\varkappa(x,t)f(x,t,v)}{\delta x}$$

$$= \frac{3}{2}\cdot\frac{c}{n}\varkappa_{max}[\varkappa_{max}-\varkappa(x,t)]\left\{\int_{v'=v}^{1}[f_m(v')f_w(x,t,v)+f_m(v)f_w(x,t,v')]dv'\right.$$

$$\left. -(1+\gamma(v(x,t)-E(V(x,t)))f(x,t,v)\right\} \qquad (II.215)$$

Hence

$$f_m(v) = \begin{cases} \dfrac{\pi}{2}\dfrac{[1+\gamma(v-2\overline{v})](\overline{v}-v)}{\overline{v}^2}\exp\left[-\dfrac{\pi}{4}\left(\dfrac{\overline{v}-v}{\overline{v}}\right)^2\right], & v>\overline{v} \\ 0, & v\leq\overline{v} \end{cases}$$

with

$$\overline{v} = \frac{c(\varkappa_{max}-\varkappa)}{2\varkappa}.$$

Taking the 0-th moment of Eq. (II.215), we obtain the well-known continuity equation

$$\frac{\delta\varkappa}{\delta t} + \frac{\delta\lambda}{\delta x} = 0.$$

This equation cannot be solved without the introduction of further information [for example, $\lambda=\lambda(\varkappa)$]. Further moments of Eq. (II.215) may be obtained. For the first moment

$$\frac{\delta v}{\delta t} + v\frac{\delta v}{\delta x} = \lambda[V(\varkappa)-v] - \frac{1}{\varkappa}\frac{\delta\varkappa\cdot var(V)}{\delta x}. \qquad (II.216)$$

In this case

$$\lambda=\lambda(\varkappa) = \frac{3}{2}\cdot\frac{c}{n}\cdot\frac{\varkappa_{max}(\varkappa_{max}-\varkappa)}{\varkappa},$$

$V(\varkappa)$ is the average speed in the steady state case, and $var(V)$ denotes the standard derivation of the speed distribution. Note the similarity between Eq. (II.216) and the acceleration equation of the continuum model [Eq. (II.206)].

By taking the first moment of the Boltzmann equation a new equation is obtained together with a new variable, the variance of the speed distribution.

The product $\varkappa\cdot var(V)$ arising in Eq. (II.216) is, by analogy with gas pressure in the kinetic theory of gases, referred to as traffic pressure P.

The first order continuum model derived from the kinetic model consists of the continuity equation, Eq. (II.216), and a function

$$P=P(\varkappa,v).$$

Assuming, for example, that traffic pressure P is independent of speed, we obtain the model

$$\frac{\delta \varkappa}{\delta t} + \frac{\delta \lambda}{\delta x} = 0$$

$$\frac{\delta v}{\delta t} + v \frac{\delta v}{\delta x} = \lambda (V (\varkappa) - v) - \frac{1}{\varkappa} \frac{dP}{d\varkappa} \frac{\delta \varkappa}{\delta x}$$

$$P = P (\varkappa)$$

where $P = P (\varkappa)$ must be determined by observation.

By constructing further moments for Eq. (II.215), one can obtain arbitrarily complicated continuum models. With each new moment, a new equation is generated together with a new variable.

References

Chapter I

General

1 Potthoff, G.: Verkehrsströmungslehre. 3. Bd. Berlin: Transpress, VEB Verlag für Verkehrswesen 1965.
2 Drew, D.R.: Traffic flow theory and control. New York: McGraw-Hill 1968.

Section I.1

3 Tölke, F.: Mechanik deformierbarer Körper. Berlin: Springer 1949.
4 Leutzbach, W.: Bewegung als Funktion von Zeit und Weg. Transportation Research 3 (1968).
5 Tournerie, G.: Sur la Définition des Grandeurs charactéristiques d'une Circulation. Straßenbau und Straßenverkehrstechnik 86 (1969) 241-244.
6 Zimmermann, W.: Zu einigen Problemen der Erhöhung der Geschwindigkeit. DDR-Verkehr 7 (1970) 283-290.

Section I.2

7 Lee, Y.W.: Statistical theory of communication. 6th edition. New York: Wiley 1967.
8 Leutzbach, W.; Steierwald, G.: Statistische und kinematische Betrachtung der Fahrt von Einzelfahrzeugen. Straßenverkehrstechnik 2 (1969) 42−45.
9 Köhler, U.: Der Zusammenhang zwischen Geschwindigkeitsganglinie bzw. Geschwindigkeitsprofil und Häufigkeitsdichte der Geschwindigkeiten. Karlsruhe: Inst für Verkehrswesen, Prel. Rep. No. 17 (1971).
10 Edie, L.C.: Flow theories. In: Gazis, D.C. (Ed.): Traffic Science. New York: Wiley 1974.
11 Winzer, Th.: Beschleunigungsverteilungen von Fahrzeugen auf zweispurigen BAB-Richtungsfahrbahnen. Straßenbau und Straßenverkehrstechnik 319 (1980).

Chapter II

General

12 Theory of traffic flow. Proc. Symp. Theory of traffic flow, Warren/Mich. 1959. Amsterdam: Elsevier 1961.
13 Haight, F.A.: Mathematical theories of traffic flow. New York: Academic Press 1963.
14 Drew, D.R.: Traffic flow theory and control. New York: McGraw-Hill 1968.
15 Gerlough, D.L.; Capelle, D.G.: An introduction to traffic flow theory. Washington: Highway Research Board, Sp. Rep. 79, 1964.
16 Vehicular traffic science. Proc. 3rd Int. Symp. Theory of traffic flow, New York 1965, New York: Elsevier 1967.
17 Beiträge zur Theorie des Verkehrsflusses. IV. Int. Symp. Theorie des Verkehrsflusses, Karlsruhe 1968. Straßenbau und Straßenverkehrstechnik 86 (1969).
18 Edie, L.C.: Flow theories. In: Gazis, D.C. (Ed.): Traffic Science. New York: Wiley 1974

19 Gerlough, D.L.; Huber, M.J.: Traffic flow theory, a monograph. Spec. Rep. 165 Transp. Res. Board, Nat. Res. Council, Washington D.C. 1975.
20 Herman, R.: Remarks on traffic flow theories and the characterization of traffic in cities. In: Proc. Workshop on "Dissipative structures in the social and physical sciences"; Univ. Texas at Austin. Austin/Texas; Univ. of Texas Press 1982.
21 Gipps, P.G. (Ed.): Traffic flow theory. Esso-Monash Series of Short Courses in Traffic Science, Clayton 1984.
22 Theorie des Verkehrsflusses auf Straßen und deren Anwendung. Forschungsges. für Straßen- und Verkehrswesen, Köln 1984.

Section II.1

23 Kreyszig, E.: Statistische Methoden und ihre Anwendungen. Göttingen: Vandenhoek und Ruprecht 1968.

Sections II.2.1 and II.2.2

24 Treiterer, J. et al.: Investigation and measurement of traffic dynamics. Appx. IX to final Report EES 202-2, Columbus: Ohio State Univ. 1965.
25 Lenz, K.-H.: Die Verkehrsmenge — Versuch einer mathematisch-statistischen Interpretation. Straßenverkehrstechnik 3/4 (1967) 31-32.
26 Jacobs, F.: Untersuchungen zur stochastischen Theorie des Verkehrsablaufs auf Straßen. Straßenbau und Straßenverkehrstechnik 96 (1970).
27 Jacobs, F.: Über die Statistik der Verkehrsstärke von Fahrzeugströmen, Forschungsgesellschaft für Straßen- und Verkehrswesen, Arbeitspapier No. 2, Köln 1984.

Section II.2.3

28 Leutzbach, W.; Egert, Ph.: Geschwindigkeitsmessungen vom fahrenden Fahrzeug aus. Straßenverkehrstechnik 3 (1959) 91-96.
29 Mori, M.; Takata, H.; Kisi, T.: Fundamental considerations on the speed distribution of road traffic flow. Transportation Research 2 (1968), 31−39.
30 Brilon, W.: Description of traffic flow by the process of slowness. Proc. 7th Int. Symp. Transportation and traffic theory, Kyoto 1977.

Section II.2.4

31 Poisson and traffic. The Eno Foundation for Highway Traffic Control, Saugatuck 1955.
32 Leutzbach, W.: Ein Beitrag zur Zeitlückenverteilung gestörter Straßenverkehrsströme. Dissertation TH Aachen 1956. Summary: International Road Safety and Traffic Reviews 3 (1957) 31-36.
33 Ferschl, F.: Zufallsabhängige Wirtschaftsprozesse — Grundlagen und Anwendungen der Theorie der Wartesysteme. Wien: Physika-Verlag 1964.
34 Leutzbach, W.; Koehler, R.: Binnenwasserstraßenverkehr als Zufallsverteilung. Karlsruhe: Institut für Verkehrswesen, Prel. Rep. No. 1, 1964.
35 Lenz, K.-H.; Garsky, J.: Anwendung mathematisch-statistischer Verfahren in der Straßenverkehrstechnik. Bad Godesberg: Kirschbaum 1968.
36 Lehmann, S.: Eine statistische Untersuchung über die Verteilung von Zeitlücken im Verkehr auf offenen Straßen. Köln: Westdeutscher Verlag 1967.

Kreyszig, E.: Statistische Methoden und ihre Anwendungen. Göttingen: Vandenhoek und Ruprecht 1968.

Section II.2.6

Leutzbach, W.; Egert, Ph.: Geschwindigkeitsmessungen vom fahrenden Fahrzeug aus. Straßenverkehrstechnik 3 (1959) 91−96.
37 Edie, L.C.: Discussion of traffic stream measurements and definitions. Proc. 2nd Int. Symp. Theory of traffic flow, London 1963. Paris: OECD 1965.

38 May, A.D.; Keller, H.E.M.: Evaluation of single − and − multi-regime traffic flow models. IV. Int. Symp. Theorie des Verkehrsflusses, Karlsruhe 1968.

39 Coers, H.G.: Die internationale Forschungsentwicklung und das räumlich-zeitliche Prinzip mikroskopischer und makroskopischer Untersuchungen des Verkehrsflusses. Die Straße 7 (1970) 368−375.

40 Dilling, J.: Charakteristik des Verkehrsablaufs auf einem Autobahnabschnitt. Karlsruhe: Institut für Verkehrswesen, Institutsnotiz No. 6, 1970.

41 Lenz, K.-H.; Ernst, R.: Untersuchungen über den Verkehrsablauf und die zulässige Geschwindigkeit auf den Behelfsfahrstreifen im Bereich der Reparaturbaustellen der Bundesautobahnen. Köln: Bundesanstalt für Straßenwesen. Prel. Report for F.A. 228/3.915, 1971.

42 Beckmann, H. et al.: Das Fundamentaldiagramm. Forschungsarbeiten aus dem Straßenwesen, Heft 89; Bad Godesberg: Kirschbaum 1973.

43 Treiterer, J.; Myers, J.: The hysterisis phenomenon in traffic flow. Proc. 6th Int. Symp. Transportation and traffic theory; Sydney: Reed 1974.

44 Leutzbach, W.; Wiedemann, R.: Traffic flow in upgrade-bottlenecks. Proc. 7th Int. Symp. Transportation and traffic theory, Kyoto 1977.

45 Hewitt, R.H.: Traffic flow theory. The Traffic Engineer 1979.

46 Leutzbach, W.: Zur Problematik der Messungen und Beobachtungen aus einem fahrenden Fahrzeug. Institut für Verkehrswesen, Universität Karlsruhe. Prel. Report No. 25, 1981.

Section II.3.1

47 Korte, J.W.; Leutzbach, W.; Mäcke, P.: Zur Frage des Überholens im Straßenverkehr. Straße und Autobahn 8 (1955), 282−284.

48 Leutzbach, W.; Egert, Ph.: Geschwindigkeitsmessungen vom fahrenden Fahrzeug aus. Straßenverkehrstechnik 3 (1959) 91−96.

49 Jacobs, F.: Untersuchungen zur stochastischen Theorie des Verkehrsablaufs auf Straßen. Straßenbau und Straßenverkehrstechnik 96, (1970).

Section II.3.2

50 Prigogine, I.: A Boltzmann-like approach to the statistical theory of traffic flow. Proc. Symp. Theory of traffic flow. Warren/Mich., 1959; New York: Elsevier 1961.

51 Munjal, P.; Pahl, J.: An analysis of the Boltzmann-type statistical models for multi-lane traffic flow. Transportation Research 3 (1969) 151−163.

52 Prigogine, I.; Herman, R.: Kinetic theory of vehicular traffic. New York: Elsevier 1971.

53 Rørbech, I.: The multilane traffic flow process. Ministry of Public Works, Road Department Copenhagen 1974.

Section II.3.2.1

54 Gebhardt, D.: Ein analytisches Warteschlangenmodell für den Verkehr auf Autobahnen. Zeitschrift für Operations Research 16 (1972) 57−61.
Rørbech, J.: The multilane traffic flow process. Ministry of Public Works, Road Department, Copenhagen 1974.

55 Brilon, W.: Warteschlangenmodell des Verkehrsablaufs auf zweispurigen Landstraßen. Straßenbau und Verkehrstechnik 201, 1976.

56 Pöschl, F.J.: Die nicht lichtsignalgeregelte Nebenstraßenzufahrt als verallgemeinertes M/G/1 Warteschlangensystem. Zeitschrift für Operations Research 27 (1983) 91−111.

Section II.3.2.2

57 Herman, R.; Lam, T.: On the mean speed in the 'Boltzmann-like' traffic theory: analytical deviation. Transportation Science 5 (1971) 314−327.

58 Gafarian, A.V.; Pahl, J.: An experimental validation of two Boltzmann-type statistical models for multilane traffic flow. Transportation Research 5 (1971) 211-224.

59 Beylich, A.E.: Untersuchungen zur kinetischen Theorie des Verkehrsflusses. Forschungsbericht No. 2662 des Landes NRW, Westdeutscher Verlag 1977.

60 Phillips, W.F.: Kinetic model for traffic flow. Rep. No. DOT/RSPD/DPB/50-77/1, Mech.
    Eng. Dept., Utah State University 1977.
61 Lampis, M.: On the kinetic theory of traffic flow in the case of a nonnegligible number of
    queueing vehicles. Transportation Science 12 (1978) 16−28.
62 Phillips, W.F.: A kinetic model for traffic flow with continuum implications. Transportation
    Planning and Technology 5 (1978) 131−138.
63 Svenson, A.: An equilibrium equation for road traffic. Transportation Research 12
    (1978) 309–313.
64 Edie, L.C.; Herman, R.: Observed multilane speed distributions and the kinetic theory of
    vehicles traffic. Transportation Science 14 (1980).
65 Beylich, A.E.; Poethke, H.J.: Gedächtnisfunktionen und Momentenverfahren in der
    kinetischen Theorie des Verkehrsflusses. Forschungsbericht No. Be802/1 der Deutschen
    Forschungsgemeinschaft, 1981.

Section II.3.2.3

66 Poethke, H.J.: Ein Vierphasenmodell des Verkehrsflusses auf Autobahnen. Dissertation an
    der Fakultät für Maschinenwesen der RWTH Aachen, 1982.

Section II.3.2.4

67 Sparmann, U.: Spurwechselvorgänge auf zweispurigen BAB-Richtungsfahrbahnen. Disser-
    tation an der Fakultät für Bauingenieur- und Vermessungswesen der Universität Karlsruhe,
    1978.
68 Leutzbach, W.; Busch, F.: Spurwechselvorgänge auf dreispurigen BAB-
    Richtungsfahrbahnen. FA. 1.082G81H des BMV, Karlsruhe 1984.
69 Sparmann, U.: Zusammenhang zwischen Geschwindigkeiten und Vorbeifahrtenhäufigkeit
    auf zweispurigen BAB-Richtungsfahrbahnen. Institut für Verkehrswesen, Universität
    Karlsruhe. Prel. Report No. 23, 1979.

Section II.3.3

70 Lighthill, M.J.; Witham, G.B.: On kinematic waves, Pt. II, A theory of traffic flow on long
    crowded roads. Proc. Royal Society, Series A, Mathematical and Physical Sciences, No.
    1178, Vol. 229, London 1955.
71 Greenberg, H.: An analysis of traffic flow. Operations Research 7 (1959) 79−85.
72 Newell, G.F.: A theory of traffic flow in tunnels. Proc. Symp. Theory of traffic flow,
    Warren/Mich. 1959. Amsterdam: Elsevier 1961.
73 Ashton, W.D.: The theory of road traffic flow. New York: Wiley, 1966.
74 Leutzbach, W.; Bexelius, S.: Probleme der Kolonnenfahrt. Straßenbau und
    Straßenverkehrstechnik 44 (1966).
75 Pipes, L.A.: Topics in the hypodynamic theory of traffic flow. Transportation Research 2
    (1968) 143−149.
76 Rockwell, T.H.; Treiterer, J.: Sensing and communication between vehicles. National
    Cooperative Highway Research Program, Report 51, Washington: HRB 1968.
77 Pipes, L.A.: Vehicle accelerations in the hydrodynamic theory of traffic flow. Transportation
    Research 3 (1969) 229−234.
    Tournerie, G.: Sur la Définition des Grandeurs charactéristiques d'une Circulation.
    Straßenbau und Straßenverkehrstechnik 86 (1969) 421, 428.
78 Haberman, R.: Mathematical models. New Jersey: Prentice Hall Inc. 1977.

Section II.3.3.1

79 Wehner, B.: Die Leistungsfähigkeit von Straßen. Berlin: Forschungsarbeiten aus dem
    Straßenwesen, Bd. 20, 1939.
80 Gazis, D.C.; Herman, R.; Potts, R.B.: Car following theory of steady state traffic flow.
    Operations Research 7 (1959) 499−505.
81 Herman, R.; Montroll, E.W.; Potts, R.B.; Rothery, R.W.: Traffic dynamics: analysis of
    stability in car following. Operations Research 7 (1959) 86−106.
82 Gazis, D.C.; Herman, R.; Rothery, R.W.: Non-linear follow-the-leader models of traffic flow.
    Operations Research 9 (1961) 545−567.

83   Wehner, B.: Der Wert von Pendelmeßwerten für die Beurteilung der Griffigkeit von Straßenoberflächen. Straße und Autobahn (1962) 458.

84   May, A.D.; Keller, H.F.M.: Non-integer car following models. Washington: HRB 199 (1967).

85   Taylor, W.J.: Traffic flow solution: graphical method. Australian Road Research 4 (1969) 77-81.

86   Hartwich, E.: Längsdynamik und Folgebewegung des Straßenfahrzeugs und ihr Einfluß auf das Verhalten der Fahrzeugschlange. Dissertation im Fachbereich 19, Regelungs- und Datentechnik TH Darmstadt, 1971.

87   Köhler, U.: Stabilitätsuntersuchungen einiger deterministischer Fahrzeugfolgegleichungen. Karlsruhe: Institut für Verkehrswesen, 1972.

88   Köhler, U.: Stabilität von Fahrzeugkolonnen. Schriftenreihe des Inst. für Verkehrswesen der Universität Karlsruhe, Heft 9, 1974.

89   Jahnke, C.D.: Kolonnenverhalten von Fahrzeugen mit autarken Abstandswarnsystemen. Schriftenreihe des Inst. für Verkehrswesen der Universtität Karlsruhe, Heft 23, 1982.

Section II.3.3.2

90   Michaels, R.M.: Perceptual factors in car following. Proc. 2nd Int. Symp. Theory of traffic flow, London 1963. Paris: OECD, 1965.

91   Todosiev, E.P.: The action-point model of the driver-vehicle-system. Columbus: The Ohio State University, Report 202 A-3, 1963.

92   Wiedemann, R.: Verkehrsablauf hinter Lichtsignalanlagen. Straßenbau und Straßenverkehrstechnik 74, 1968.

93   Wiedemann, R.: Simulation des Verkehrsflusses. Schriftenreihe des Instituts für Verkehrswesen der Universität Karlsruhe, Heft 8, 1974.

94   Hoefs, D.H.: Untersuchung des Fahrverhaltens in Fahrzeugkolonnen. Straßenbau und Straßenverkehrstechnik 140 (1972).

Section II.3.3.3

     Jacobs, F.: Untersuchungen zur stochastischen Theorie des Verkehrsablaufs auf Straßen. Straßenbau und Straßenverkehrstechnik 96, 1970.

95   Treiterer, J. et al.: Investigation of traffic dynamics by aerial photogrammetric techniques. Interim Report EEs 278-3, Columbus: Ohio State University, 1970.

96   Munjal, P.K. et al.: Analysis and validation of lane-drop effects on multi lane freeways. Transportation Research 5 (1971) 257−266.

97   Munjal, P.K.; Pipes, L.A.: Propagation of on-ramp density waves on uniform unidirectional multilane freeways. Transportation Science 5 (1971) 390−402.

98   Munjal, P.K.; Pipes, L.A.: Propagation of on-ramp density perturbations on unidirectional two- and three-lane freeways. Transportation Research 5 (1971) 241−255.

99   Willmann, G.: Stauberechnung als Entscheidungshilfe bei Verkehrslenkungsmaßnahmen. 10 Jahre Institut für Verkehrswesen, Schriftenreihe des Instituts für Verkehrswesen der Universität Karlsruhe, Heft 6, 1972.

100 Leutzbach, W.; Köhler, U.: Definitions and relationships for three different time intervals for delayed vehicles. Proc. 6th Int. Symp. Transportation and traffic theory, Sydney: Reed 1974.

101 Phillips, W.F.: A kinetic model for traffic flow with continuum implications. Transportation Planning and Technology 5 (1978) 131−138.

102 Stephanopoulos, G. et al.: Modelling and analysis of traffic queue dynamics at signalized intersections. Transportation Research 13A (1979) 295−307.

103 Michalopoulos, P.G. et al.: An application of shockwave theory to traffic signal control. Transportation Research 15B (1981) 35−51.

104 Michalopoulos, P.G.; Beskos, D.E.: Improved continuum models of freeway flow. Proc. 9th Int. Symp. Transportation and traffic theory. VNU Science Press, Utrecht 1984.

105 Kühne, R.: Macroscopic freeway model for dense traffic-stop-start waves and incident detection. Proc. 9th Int. Symp. Transportation and traffic theory, VNU Science Press, Utrecht 1984.

106 Kühne, R.: Fernstraßenverkehrsbeeinflussung und Physik der Phasenübergänge. Physik in unserer Zeit 3, 1984.
107 Sasaki, T. et al.: An approximative analysis of the hydrodynamic theory on traffic flow and a formulation of a traffic simulation model. Proc. 9th Int. Symp. Transportation and traffic theory, VNU Science Press, Utrecht, 1984.

List of Symbols

e	Euler's number $e = 2,7182818$
ln	natural logarithm
$P(i)$	probability for event i
$E(i)$	expected value for continuous random variable i
$x(t)$	distance as a function of time
$v(t), u(t)$	speed as a function of time
v_{min}	minimum speed
v_{max}	maximum speed
t	time
$b(t)$	acceleration as a function of time
$k(t)$	jerk as a function of time
t_0, x_0, v_0, b_0	initial conditions
v_A, t_A	index A means beginning
v_E, b_E	index E means ending
x_R, t_R	index R means reaction
x_B, t_B	index B means braking
t_v	time as a function of speed
x_v	distance as a function of speed
$w(x)$	slowness as a function of distance
$c(x)$	change of w as a function of distance
$l(x)$	change of c as a function of distance
$f_t(v)$	relative frequency as a function of speed
$F_t(v)$	relative cumulative frequency of speed
v_i	classes of speed
$\bar{v}$	arithmetic mean of speed
s^2	variance, discrete
s	standard deviation, discrete
σ^2	variance
σ	standard deviation
$\bar{\sigma}^2$	variance estimated from probability density function
∂	skewness or asymmetry
$\hat{v}_t$	mean value of speed-time-profile
$\hat{v}_x$	mean value of speed-distance-profile
$\hat{w}_x$	mean value of slowness-distance-profile

$\hat{w}_t$	mean value of slowness-time-profile
X, L	distance as a difference between two points
$t(x), v(x), b(x)$	time, speed, acceleration as a function of distance
$t(v), x(v), b(v)$	time, distance, acceleration as a function of speed
k	traffic density
λ_{x_i}	intensity of the traffic stream at point x_i
q	traffic volume
$\Phi_{x_i}(t)$	CUSUM function: stream function at observation point x_i as a function of time
M	vehicle count at x_i in time interval Δt
N	number of vehicles at t_i in Δx
$\varkappa_{t_i}$	concentration at time t_i
$\Psi_{t_i}(x)$	CUSUM function: stream function at observation point t_i as a function of distance
$n(x,t)$ $n'(x,t)$	traffic stream as a function of time and distance
ACN	acceleration noise
l	index l means local
m	index m means instantaneous
$G_m(v)$	speed distribution from instantaneous measurement
$g_m(v)$	probability density function corresponding to $G_m(v)$
$G_l(v)$	speed distribution from local measurement
$g_l(v)$	probabiliy density function corresponding to $G_1(v)$
l_f	vehicle length
z_i	time headways
a_i	distance headways
q_p^p	passive overtakings per time-interval
$q_p^{a\cdot}$	active overtakings per time-interval
M_p^p	passive overtakings, total number
M_p^a	active overtakings, total number

v_w desired speed R_p^p number of passive overtaking
$\bar{v}_w$ mean desired speed   per unit-time and distance of all
r travel time   vehicles
R_p^a number of active overtakings c speed of kinematic waves
                 per unit-time and distance of all
                 vehicles

Subject Index

Definitions of entries will be given at page numbers printed in *italic*.